Springer Proceedings in Complexity

Springer Complexity

Springer Complexity is an interdisciplinary program publishing the best research and academic-level teaching on both fundamental and applied aspects of complex systems—cutting across all traditional disciplines of the natural and life sciences, engineering, economics, medicine, neuroscience, social, and computer science.

Complex Systems are systems that comprise many interacting parts with the ability to generate a new quality of macroscopic collective behavior the manifestations of which are the spontaneous formation of distinctive temporal, spatial, or functional structures. Models of such systems can be successfully mapped onto quite diverse "real-life" situations like the climate, the coherent emission of light from lasers, chemical reaction—diffusion systems, biological cellular networks, the dynamics of stock markets and of the Internet, earthquake statistics and prediction, freeway traffic, the human brain, or the formation of opinions in social systems, to name just some of the popular applications.

Although their scope and methodologies overlap somewhat, one can distinguish the following main concepts and tools: self-organization, nonlinear dynamics, synergetics, turbulence, dynamical systems, catastrophes, instabilities, stochastic processes, chaos, graphs and networks, cellular automata, adaptive systems, genetic algorithms, and computational intelligence.

The three major book publication platforms of the Springer Complexity program are the monograph series "Understanding Complex Systems" focusing on the various applications of complexity, the "Springer Series in Synergetics", which is devoted to the quantitative theoretical and methodological foundations, and the "SpringerBriefs in Complexity" which are concise and topical working reports, case-studies, surveys, essays, and lecture notes of relevance to the field. In addition to the books in these two core series, the program also incorporates individual titles ranging from textbooks to major reference works.

More information about this series at http://www.springer.com/series/11637

Erez Shmueli • Baruch Barzel • Rami Puzis
Editors

3rd International Winter School and Conference on Network Science

NetSci-X 2017

 Springer

Editors
Erez Shmueli
Department of Industrial Engineering
Tel Aviv University
Tel Aviv, Israel

Baruch Barzel
Department of Mathematics
Bar-Ilan University
Ramat-Gan, Israel

Rami Puzis
Department of Software
    and Information Systems Engineering
Ben-Gurion University of the Negev
Beer-Sheva, Israel

ISSN 2213-8684   ISSN 2213-8692   (electronic)
Springer Proceedings in Complexity
ISBN 978-3-319-85671-1   ISBN 978-3-319-55471-6   (eBook)
DOI 10.1007/978-3-319-55471-6

Printed on acid-free paper

This Springer imprint is published by Springer Nature
The registered company is Springer International Publishing AG
The registered company address is: Gewerbestrasse 11, 6330 Cham, Switzerland

Contents

Node-Centric Detection of Overlapping Communities in Social Networks

Yehonatan Cohen, Danny Hendler, and Amir Rubin

Abstract We present NECTAR, a community detection algorithm that generalizes Louvain method's local search heuristic for overlapping community structures. NECTAR chooses dynamically which objective function to optimize based on the network on which it is invoked. Our experimental evaluation on both synthetic benchmark graphs and real-world networks, based on ground-truth communities, shows that NECTAR provides excellent results as compared with state of the art community detection algorithms.

Keywords Community detection • Overlapping communities • Extended modularity • Louvain method • Weighted community clustering

1 Introduction

Social networks tend to exhibit community structure [1], that is, they may be partitioned to sets of nodes called *communities* (a.k.a. *clusters*), each of which relatively densely-interconnected, with relatively few connections between different communities. Revealing the community structure underlying complex networks in general, and social networks in particular, is a key problem with many applications (see e.g. [2, 3]) that is the focus of intense research. Numerous community detection algorithms were proposed (see e.g. [4–14]). While research focus was initially on detecting *disjoint communities*, in recent years there is growing interest in the detection of *overlapping communities*, where a node may belong to several communities.

Many community detection algorithms are guided by an *objective function* that provides a quality measure of the clusterings they examine in the course of their execution. Since exhaustive-search optimization of these functions is generally intractable (see e.g. [15, 16]), existing methods settle for an approximation of the optimum and employ heuristic search strategies.

Y. Cohen • D. Hendler • A. Rubin (✉)
Computer Science Department, Ben-Gurion University of the Negev, Be'er-Sheva, Israel
e-mail: yehonatc@cs.bgu.ac.il; hendlerd@cs.bgu.ac.il; amirrub@cs.bgu.ac.il

© Springer International Publishing AG 2017
E. Shmueli et al. (eds.), *3rd International Winter School and Conference on Network Science*, Springer Proceedings in Complexity, DOI 10.1007/978-3-319-55471-6_1

A key example is Blondel et al.'s algorithm [8], also known by the name "Louvain method" (LM). The algorithm is fast and relatively simple to understand and use and has been successfully applied for detecting communities in numerous networks. It aims to maximize the modularity objective function [9]. Underlying it is a greedy local search heuristic that iterates over all nodes, assigning each node to the community it fits most (as quantified by modularity) and seeking a local optimum. Unfortunately, the applicability of LM is limited to disjoint community detection.

Our Contributions We present NECTAR, a Node-centric ovErlapping Community deTection AlgoRithm. NECTAR generalizes the node-centric local search heuristic of the Louvain algorithm so that it can be applied also to networks possessing overlapping community structure. Several algorithmic issues have to be dealt with in order to allow the LM heuristic to support multiple community-memberships per node. First, rather than adding a node v to the *single* community maximizing an objective function, v may have to be added to several such communities. However, since the "correct" number of communities to which v should belong is not a-priori known to the algorithm, it must be chosen dynamically.

A second issue that arises from multiple community-memberships is that different communities with large overlaps may emerge during the algorithm's execution and must be merged. We describe the new algorithm and how it resolves these issues in Sect. 2.

Modularity (used by LM) assumes disjoint communities. Which objective functions should be used for overlapping community detection? Yang and Leskovec [17] evaluated several objective functions and showed that which is most appropriate depends on the network at hand. They observe that objective functions that are based on triadic closure provide the best results when there is significant overlap between communities. Weighted Community Clustering (WCC) [18] is such an objective function but is defined only for disjoint community structures.

We define Weighted Overlapping Community Clustering (WOCC), a generalization of WCC that may be applied for overlapping community detection. More details can be found in our technical report [19]. Another objective function that fits the overlapping setting is Q^E [20]—an extension of modularity for overlapping communities.

A unique feature of NECTAR is that it chooses dynamically whether to use WOCC or Q^E, depending on the structure of the graph at hand. This allows it to provide good results on graphs with both high and low community overlaps. NECTAR is the first community-detection algorithm that selects dynamically which objective function to use based on the graph on which it is invoked.

Local search heuristics guided by an objective function may be categorized as either *node-centric* or *community-centric*. Node-centric heuristics iterate over nodes. For each node, communities are considered and it is added to those of them that are "best" in terms of the objective function. Community-centric heuristics do the opposite: they iterate over communities. For each community, nodes are considered and the "best" nodes are added to it. In order to investigate which of these approaches is superior in the context of social networks, we implemented both a node-centric and a community-centric versions of NECTAR and compared the two implementations using both the WOCC and the Q^E metrics. As can be seen in our technical report[19], the node-centric approach was significantly superior for both metrics used.

We conducted extensive competitive analysis of NECTAR (using a node-centric approach) and nine other state-of-the-art overlapping community detection algorithms. Our evaluation was done using both synthetic graphs and real-world networks with ground-truth communities, based on several commonly-used metrics. NECTAR outperformed all other algorithms in terms of average detection quality and was best or second-best for almost all networks. Our code is publicly available for download.[1]

Background We now briefly describe a few key notions directly related to our work. Louvain method [8] is a widely-used disjoint community detection algorithm, based on a simple node-centric search heuristic that seeks to maximize the *modularity* [9] objective function. Chen et al. extended the definition of modularity to the overlapping setting [20]. For a collection of sets of nodes $\mathscr{C}$, their *extended modularity* definition, denoted $Q^E(\mathscr{C})$, is given by:

$$Q^E(\mathscr{C}) = \frac{1}{2|E|} \sum_{C \in \mathscr{C}} \sum_{i,j \in C} \left[A_{ij} - \frac{k_i k_j}{2|E|} \right] \frac{1}{O_i O_j}, \tag{1}$$

where A is the adjacency matrix, k_i is the degree of node i, and O_i is the number of communities i is a member of. If $\mathscr{C}$ is a partition of network nodes, Q^E reduces to (regular) modularity.

Yang and Leskovec [17] conducted a comparative analysis of 13 objective functions in order to determine which captures better the community structure of a network. They show that which function is best depends on the network at hand. They also observe that objective functions that are based on *triadic closure* provide the best results when there is significant overlap between communities.

Weighted Community Clustering (WCC) [18] is such an objective function. It is based on the observation that triangle structures are much more likely to exist within communities than across them. This observation is leveraged for quantifying the quality of graph partitions (that is, non-overlapping communities). It is formally defined as follows. For a set of nodes S and a node v, let $t(v, S)$ denote the number of triangles that v closes with nodes of S. Also, let $vt(v, S)$ denote the number of nodes of S that form at least one triangle with v. $WCC(v, S)$, quantifying the extent by which v should be a member of S, is defined as:

$$WCC(v, S) = \begin{cases} \frac{t(v,S)}{t(v,V)} \cdot \frac{vt(v,V)}{|S \backslash v| + vt(v, V \backslash S)} & \text{if } t(v, V) > 0 \\ 0 & \text{otherwise,} \end{cases}$$

[1]NECTAR code and documentation may be downloaded from: https://github.com/amirubin87/NECTAR.

where V is the set of graph nodes. The cohesion level of a community S is defined as $WCC(S) = \frac{1}{|S|} \sum_{v \in S} WCC(v, S)$. Finally, the quality of a partition $\mathscr{C} = \{S_1, \ldots, S_k\}$ is defined as the following weighted average: $WCC(\mathscr{C}) = \frac{1}{|V|} \sum_{i=1}^{k} |S_i| \cdot WCC(S_i)$.

NECTAR uses Weighted Overlapping Community Clustering (WOCC)—our generalization of WCC that can be applied to overlapping community detection.

2 NECTAR: A Detailed Description

The high-level pseudo-code of NECTAR is given by Algorithm 1. The input to the NECTAR procedure (see 4) is a graph $G = <V, E>$ and an algorithm parameter $\beta \geq 1$ that is used to determined the number of communities to which a node should belong in a dynamic manner (as we describe below).

NECTAR proceeds in iterations (lines **12–28**), which we call *external iterations*. In each external iteration, the algorithm performs *internal iterations*, in which it iterates over all nodes $v \in V$ (in some random order), attempting to determine the set of communities to which node v belongs such that the objective function is maximized.

We implemented two overlapping community objective functions: the extended modularity function [20], denoted $Q^E(\mathscr{C})$, and WOCC—our generalization of the WCC function [18]. These implementations are described in our technical report [19, 21]. NECTAR selects dynamically whether to use WOCC or Q^E, depending on the rate of closed triangles in the graph on which it is invoked. If the average number of closed triangles per node in G is above the *trRate* threshold, then WOCC is more likely to yield good performance and it is used, otherwise the extended modularity objective function is used instead (lines **5–8**). We use *trRate* $= 5$, as this provides a good separation between communities with high overlap (on which WOCC is superior) and low overlap (on which extended modularity is superior).

Each internal iteration (comprising lines **13–23**) proceeds as follows. First, NECTAR computes the set C_v of communities to which node v currently belongs (line **14**). Then, v is removed from all these communities (line **15**). Next, the set S_v of v's neighboring communities (that is, the communities of $\mathscr{C}$ that contain one or more neighbors of v) is computed in line **16**. Then, the gain in the objective function value that would result from adding v to each neighboring community (relative to the current set of communities $\mathscr{C}$) is computed in line **17**. Node v is then added to the community maximizing the gain in objective function and to any community for which the gain is at least a fraction of $1/\beta$ of that maximum (lines **18–19**).[2] Thus, the number of communities to which a node belongs may change dynamically throughout the computation, as does the set of communities $\mathscr{C}$.

---

[2]If no gain is positive, v remains as a singleton.

If the internal iteration did not change the set of communities to which v belongs, then v is a *stable node* of the current external iteration and the number of stable nodes (initialized to 0 in line **13**) is incremented (lines **20–21**).

After all nodes have been considered, the possibly-new set of communities is checked in order to prevent the emergence of different communities that are too similar to one another. This is done by the `merge` procedure (whose code is not shown), called in line **24**. It receives as its single parameter a value α and merges any two communities whose relative overlap is α or more. If the number of communities was reduced by `merge`, the counter of stable nodes is reset to 0 (lines **25–26**).

The computation proceeds until either the last external iteration does not cause any changes (hence the number of stable nodes equals $|V|$) or until the maximum number of iterations is reached (line **28**), whichever occurs first. We have set the maximum number of iterations to 20 (line **1**) in order to strike a good balance between detection quality and runtime. In practice, the algorithm converges within a fewer number of iterations in the vast majority of cases. For example, in our experiments on synthetic graphs with 5000 nodes, NECTAR converges after at most 20 iterations in 99.5% of the executions.

LM is a hierarchical clustering algorithm that has a second phase. We implemented a hierarchical version of NECTAR. However, since in all our experiments the best results were obtained in the first hierarchy level, we only describe the non-hierarchical version of NECTAR (Algorithm 1).

3 Experimental Evaluation

Xie et al. [22] conducted a comparative study of state-of-the-art overlapping community detection algorithms. We compare NECTAR with the following 5 of

---

Figure 1: NECTAR algorithm

```
1  const maxIter ← 20                          12  repeat
2  const α ← 0.8                               13  |   s ← 0 forall the v ∈ V do
3  const trRate ← 5                            14  |   |   C_v ← v's communities
                                               15  |   |   Remove v from all C_v communities
4  Procedure NECTAR(G=<V,E>, β){               16  |   |   S_v ← {C ∈ 𝒞 | ∃u : u ∈ C ∧ (v, u) ∈ E}
5  if triangles(G)/|V| ≥ trRate  then          17  |   |   D ← {Δ(v, C) | C ∈ S_v}
6  |   use WOCC                                18  |   |   C'_v ← {C ∈ S_v | Δ(v, C) · β ≥ max(D)}
7  else                                        19  |   |   Add v to all the communities of C'_v
8  |   use Q^E                                 20  |   |   if C'_v = C_v  then
9  end                                         21  |   |   |   s++
10  Initialize communities                     22
11  i ← 0                                       23  |   end
                                               24  |   merge(α)
                                               25  |   if merge reduced communities num.  then
                                               26  |   |   s ← 0
                                               27  |   i++
                                               28  until (s = |V|) ∨ (i = maxIter)
```

the key performers out of the 14 algorithms they evaluated: the *Greedy Clique Expansion* (GCE) algorithm [23], the *Cfinder* algorithm [12], the *Order Statistics Local Optimization Method* (OSLOM) [13], the *Community Overlap PRopagation Algorithm* (COPRA) [24], and the *Speaker-Listener Label Propagation Algorithm* (SLPA) [10]. In addition, we also evaluate the following four algorithms: *Fuzzy-Infomap* [14], *Big-Clam* [25], *Link-Clustering* (LC) [26], and *DEMON* [27]. Details regarding these algorithms and the parameters we used when invoking them can be found in our technical report [19, 21].

We conducted competitive analysis using both synthetic networks and real-world networks with ground-truth. We evaluated results using the widely-used *Normalized Mutual Information* (NMI) [5], *Omega-index* [28], and *Average F1 score* [29] metrics (descriptions of these metrics can be found in [21]). Our evaluation shows that NECTAR outperformed all other algorithms in terms of average detection quality and provided best or second-best results for almost all networks, as we describe now.

Synthetic Networks Lancichinetti et al. [30] introduced a set of benchmark graphs (henceforth the LFR benchmark), parameterized on: the number of nodes, n, the average node degree, k, the number of overlapping nodes, O_n, the number of communities an overlapping node belongs to, O_m, community sizes (varied in our experiments between 20 and 100 for big communities and between 10 and 50 for small communities), and more. We mostly use the LFR parameter values used by [22].[3] We generate 10 instances for each combination of parameters and take the average of the results for each algorithm and each metric over these 10 instances. For each algorithm, we present the results for the algorithm parameter value that maximizes this average.

Figure 1 presents the average performance of the four best algorithms in terms of NMI as a function of O_m (the number of communities to which each of the O_n overlapping nodes belongs), for $k \in \{10, 40\}$ and $O_n \in \{2500, 5000\}$. The Omega-index and average-F1 score results follow the same trends and are thus omitted for lack of space. They can be found in our technical report [19].

With only a few exceptions, it can be seen that the performance of the algorithms decreases as O_m increases. This can be attributed to the fact that the size of the solution space increases with O_m.

We focus first on the results on graphs with a higher number of overlapping nodes ($O_m = 2500$) and high average degrees ($k = 40$). The rate of triangles in these graphs is high (approx. 30 on average) and so NECTAR employs WOCC. NECTAR is the clear leader for big communities. It achieves the best results for almost all values of O_m and its relative performance improves as O_m increases, confirming that the combination of NECTAR's search strategy and the WOCC objective function is suitable for graphs with significant overlap. Cfinder improves its relative performance as O_m increases and is the second performer for $O_m \in$

---

[3]For more details on parameter values used for LFR, refer to [21].

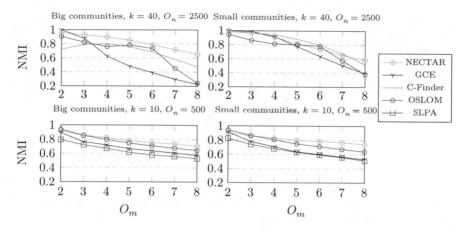

Fig. 1 Four best performers over synthetic networks in terms of NMI

$\{4, 7, 8\}$. For small communities, Cfinder has the lead with NECTAR being second best and OSLOM third for most values of O_m, and NECTAR taking the lead for $O_m = 8$.

We now describe the results on graphs with lower numbers of overlapping nodes $(O_m = 500)$ and low average degree $(k = 10)$. The rate of triangles in these graphs is low (approx. 3.5 on average) and so NECTAR employs extended modularity. NECTAR provides the best performance for both small and large communities for almost all values of O_m. The relative performance of Cfinder deteriorates as compared with its performance on high-overlap graphs. It is not optimized for sparser graphs, since its search for communities is based on locating cliques. OSLOM is second best on these graphs, having the upper hand for $O_m = 1$ and providing second-best performance for $O_m > 1$. These results highlight the advantage of NECTAR's capability of selecting the objective function it uses dynamically according to the properties of the graph at hand.

Summarizing the results of the tests we conducted on 96 different synthetic graph types, NECTAR is ranked first, with average rank of 1.58, leading in 33 out of 96 of the tests, followed by OSLOM, with average rank of 2.79.

Real-World Networks We conducted our competitive analysis on two real-world networks—Amazon's product co-purchasing network and the DBLP scientific collaboration network. We downloaded both from Stanford's Large Network Dataset Collection [31]. The Amazon graph consists of 334,863 nodes and 925,872 edges. Nodes represent products and edges are between commonly co-purchased products. Products from the same category are viewed as a ground-truth community.

The DBLP graph consists of 317,080 nodes and 1,049,866 edges. Nodes correspond to authors and edges connect authors that have co-authored a paper. Publication venues (specifically, conferences) are used for defining ground-truth

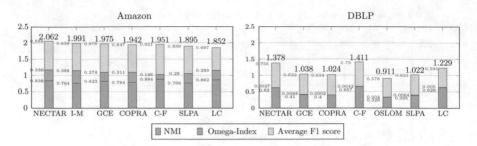

Fig. 2 Seven best performers over real-world networks

communities. Thus, the set of authors that have published in the same conference is viewed as a ground-truth community.

In [17], Yang and Leskovec rate the quality of ground-truth communities of Amazon and DBLP (as well as those of additional networks) using six scoring functions, such as modularity, conductance, and cut ratio. They rank ground-truth communities based on the average of their ranks over the six corresponding scores and maintain the 5000 top ground-truth communities per each network. These are the ground-truth communities provided as part of the datasets of [31].

The left part of Fig. 2 presents the results of the seven best algorithms on Amazon. The right part refers to results on DBLP. The rate of triangles in the Amazon graph is low, and so NECTAR employs extended modularity. NECTAR provides the best performance with an overall score of 2.062, approximately 3.5% more than InfoMap, which is second best. NECTAR has second-best average F1 score, lagging only slightly behind Cfinder. In terms of Omega-index, NECTAR is second-best as well, lagging behind InfoMap, and Cfinder is the last performer.

In the DBLP network, the rate of triangles is high, and so NECTAR employs WOCC. Cfinder has the highest overall score, enjoying a small margin of approximately 2.5% w.r.t. NECTAR, which is second-best. LC is the third performer, with a score lower than NECTAR's by approximately 8%. In terms of NMI, Cfinder is first with a score of 0.657 and NECTAR is third best, lagging behind by approximately 5.5%. NECTAR has the highest average F1 score, but Cfinder's score is only approximately 1% smaller. COPRA obtains the third score, nearly 17% less than NECTAR's. All algorithms fair poorly in terms of their Omega-index.

In order to assess the impact of dynamic objective function selection, we compared NECTAR with two variants that consistently used either Q^E or WOCC. In cases of disagreement, NECTAR's score was, on average, 30% higher than that of the WOCC version and 13% higher than the Q^E version.

We also measured time complexity on numerous networks, while varying the number of nodes and the average node degree. NECTAR's average running time was second best among all evaluated algorithms.

4 Conclusions

We introduced NECTAR, a novel overlapping community detection algorithm that generalizes Louvain's search heuristic and selects dynamically which objective function to optimize, depending on the structure of the graph at hand.

Our evaluation shows that NECTAR outperforms all other algorithms in terms of average detection quality. Analysis of our empirical results shows that extended modularity yields better results on networks with low average node degrees and low community overlap, whereas WOCC yields better results on networks with higher degrees and overlap. The fact that NECTAR is able to provide excellent results on both types of networks highlights the importance of objective function dynamic selection, as well as the general applicability of Louvain's search heuristic.

Acknowledgements Partially supported by the Cyber Security Research Center at Ben-Gurion University and by the Lynne and William Frankel Center for Computer Science.

References

1. Fortunato, S.: Community detection in graphs. Phys. Rep. **486**(3), 75–174 (2010)
2. Krogan, N.J., Cagney, G., Yu, H., Zhong, G., Guo, X., Ignatchenko, A., Li, J., Pu, S., Datta, N., Tikuisis, A.P., et al.: Global landscape of protein complexes in the yeast Saccharomyces cerevisiae. Nature **440**(7084), 637–643 (2006)
3. Flake, G.W., Lawrence, S., Giles, C.L., Coetzee, F.M.: Self-organization and identification of web communities. Computer **35**(3), 66–70 (2002)
4. Le Martelot, E., Hankin, C.: Fast multi-scale detection of relevant communities in large-scale networks. Comput. J. **56**(9), 1136–1150, p. bxt002 (2013)
5. Lancichinetti, A., Fortunato, S., Kertész, J.: Detecting the overlapping and hierarchical community structure in complex networks. New J. Phys. **11**(3), 033015 (2009)
6. Prat-Pérez, A., Dominguez-Sal, D., Larriba-Pey, J.-L.: High quality, scalable and parallel community detection for large real graphs. In: Proceedings of the 23rd International Conference on World Wide Web, pp. 225–236. ACM, New York (2014)
7. Esquivel, A.V., Rosvall, M.: Compression of flow can reveal overlapping-module organization in networks. Phys. Rev. X **1**(2), 021025 (2011)
8. Blondel, V.D., Guillaume, J.-L., Lambiotte, R., Lefebvre, E.: Fast unfolding of communities in large networks. J. Stat. Mech. Theory Exp. **2008**(10), P10008 (2008)
9. Newman, M.E., Girvan, M.: Finding and evaluating community structure in networks. Phys. Rev. E **69**(2), 026113 (2004)
10. Xie, J., Szymanski, B.K.: Towards linear time overlapping community detection in social networks. In: Advances in Knowledge Discovery and Data Mining, pp. 25–36. Springer, Berlin (2012)
11. Gregory, S.: Finding overlapping communities in networks by label propagation. New J. Phys. **12**(10), 103018 (2010)
12. Adamcsek, B., Palla, G., Farkas, I.J., Derényi, I., Vicsek, T.: Cfinder: locating cliques and overlapping modules in biological networks. Bioinformatics **22**(8), 1021–1023 (2006)
13. Lancichinetti, A., Radicchi, F., Ramasco, J.J., Fortunato, S., et al.: Finding statistically significant communities in networks. PLoS One **6**(4), e18961 (2011)

14. Gregory, S.: Fuzzy overlapping communities in networks. J. Stat. Mech. Theory Exp. **2011**(02), P02017 (2011)
15. Brandes, U., Delling, D., Gaertler, M., Görke, R., Hoefer, M., Nikoloski, Z., Wagner, D.: On finding graph clusterings with maximum modularity. In: Graph-Theoretic Concepts in Computer Science, 33rd International Workshop, WG 2007, Dornburg, Germany, 21–23 June 2007. Revised Papers, pp. 121–132 (2007)
16. Síma, J., Schaeffer, S.E.: On the np-completeness of some graph cluster measures. In: Proceedings of the SOFSEM 2006: Theory and Practice of Computer Science, 32nd Conference on Current Trends in Theory and Practice of Computer Science, Merín, Czech Republic, 21–27 January 2006, pp. 530–537 (2006)
17. Yang, J., Leskovec, J.: Defining and evaluating network communities based on ground-truth. Knowl. Inf. Syst. **42**(1), 181–213 (2015)
18. Prat-Pérez, A., Dominguez-Sal, D., Brunat, J.M., Larriba-Pey, J.-L.: Shaping communities out of triangles. In: Proceedings of the 21st ACM International Conference on Information and Knowledge Management, pp. 1677–1681. ACM, New York (2012)
19. Cohen, Y., Hendler, D., Rubin, A.: NECTAR—Technical Report, https://github.com/amirubin87/NECTAR/blob/master/Technical-Report-NECTAR.pdf (2016).
20. Chen, M., Kuzmin, K., Szymanski, B.K.: Extension of modularity density for overlapping community structure. In: 2014 IEEE/ACM International Conference on Advances in Social Networks Analysis and Mining (ASONAM), pp. 856–863. IEEE, Piscataway (2014)
21. Cohen, Y., Hendler, D., Rubin, A.: Node-centric detection of overlapping communities in social networks. arXiv preprint arXiv:1607.01683 (2016)
22. Xie, J., Kelley, S., Szymanski, B.K.: Overlapping community detection in networks: the state-of-the-art and comparative study. ACM Comput. Surv. **45**(4), 43 (2013)
23. Lee, C., Reid, F., McDaid, A., Hurley, N.: Detecting highly overlapping community structure by greedy clique expansion. arXiv preprint arXiv:1002.1827 (2010)
24. Raghavan, U.N., Albert, R., Kumara, S.: Near linear time algorithm to detect community structures in large-scale networks. Phys. Rev. E **76**(3), 036106 (2007)
25. Yang, J., Leskovec, J.: Overlapping community detection at scale: a nonnegative matrix factorization approach. In: Proceedings of the Sixth ACM International Conference on Web Search and Data Mining, pp. 587–596. ACM, New York (2013)
26. Ahn, Y.-Y., Bagrow, J.P., Lehmann, S.: Link communities reveal multiscale complexity in networks. Nature **466**(7307), 761–764 (2010)
27. Coscia, M., Rossetti, G., Giannotti, F., Pedreschi, D.: Demon: a local-first discovery method for overlapping communities. In: Proceedings of the 18th ACM SIGKDD International Conference on Knowledge Discovery and Data Mining, pp. 615–623. ACM, New York (2012)
28. Collins, L.M., Dent, C.W.: Omega: a general formulation of the rand index of cluster recovery suitable for non-disjoint solutions. Multivar. Behav. Res. **23**(2), 231–242 (1988)
29. Yang, J., Leskovec, J.: Community-affiliation graph model for overlapping network community detection. In: 2012 IEEE 12th International Conference on Data Mining (ICDM), pp. 1170–1175. IEEE, Piscataway (2012)
30. Lancichinetti, A., Fortunato, S., Radicchi, F.: Benchmark graphs for testing community detection algorithms. Phys. Rev. E **78**(4), 046110 (2008)
31. Leskovec, J., Krevl, A.: SNAP Datasets: Stanford large network dataset collection. http://snap.stanford.edu/data (2014)

Community Structures Evaluation in Complex Networks: A Descriptive Approach

Vinh-Loc Dao, Cécile Bothorel, and Philippe Lenca

Abstract Evaluating a network partition just only via conventional quality metrics—such as modularity, conductance or normalized mutual of information—is usually insufficient. Indeed, global quality scores of a network partition or its clusters do not provide many ideas about their structural characteristics. Furthermore, quality metrics often fail to reach an agreement especially in networks whose modular structures are not very obvious. Evaluating the goodness of network partitions in function of desired structural properties is still a challenge.

Here, we propose a methodology that allows one to expose structural information of clusters in a network partition in a comprehensive way, thus eventually helps one to compare communities identified by different community detection methods. This descriptive approach also helps to clarify the composition of communities in real-world networks. The methodology hence bring us a step closer to the understanding of modular structures in complex networks.

Keywords Community structure • Quality function • Community evaluation • Community detection

1 Introduction

Modular structures have been noticed in a large range of real-world networks through many researches on social networks [6, 7, 11], computer networks such as the Internet [5, 14], biochemical networks [9, 12], etc. Nodes in networks have a

V.-L. Dao (✉) • C. Bothorel • P. Lenca
Institut Mines Telecom, IMT Atlantique, UMR CNRS 6285 Lab-STICC, Brest, France
e-mail: vinh.dao@imt-atlantique.fr; cecile.bothorel@imt-atlantique.fr;
philippe.lenca@imt-atlantique.fr

© Springer International Publishing AG 2017
E. Shmueli et al. (eds.), *3rd International Winter School and Conference on Network Science*, Springer Proceedings in Complexity, DOI 10.1007/978-3-319-55471-6_2

tendency to connect preferably with the similar ones to establish functional groups, sometimes called clusters, modules or communities. Understanding modular structures of networks pays an essential role in the study of their functionalities.

Since the notion of community varies according to specific contexts, it seems not appropriate to use a global quality criteria in order to evaluate graph partitions. Depending on which kind of network is considered in which kind of application, one might need to decompose a network into clusters that possess specific features with desired structures. Once a network partition is available, the clusters need to be analyzed to verify the existence of features as well as their quality in the global image of the network.

In small networks, clusters can be evaluated manually by simple visualizations, however when the sizes grow, manual evaluation is not feasible. In these cases, expected concepts of community are mathematically translated into quality metrics such as conductance or modularity Q [3, 7, 11] in order to quantify the quality of clusters. Those quality functions score the goodness of clusters according to their associated concepts of community but can not identify or describe more specific structural patterns. In other words, many interested structure features in communities are invisible to quality functions.

In this work, we propose a methodology to describe communities through intra-cluster links and inter-cluster links in such a way that structural information is exposed comprehensively to evaluators. Such a description will help one to evaluate network partitions according to different concepts of community and to detect more sophisticated structures. Our results show that ground-truth communities composition in many real-world networks exposes a diversity in structural patterns, which are very different from the conventional notion of community.

2 Related Works

Many researches have been conducted in order to understand the nature of ground-truth communities in real-world networks as well as ones identified by community detection algorithms over a broad range of networks. Although the notion of community is not straight forward, these researches provide essential information so that one can study several qualities of communities as well as their characteristics.

Leskovec et al. [13] compared the performance of 13 quality functions in term of their efficiencies to identify community goodness properties such as density, cohesiveness. Besides, the authors also analyzed the consistence of these quality functions' performances to many simulated perturbations.

Due to the fact that community structures may strongly differ from networks to networks. Creusefond et al. [4] proposed a methodology to identify groups of networks where quality functions perform consistently. The authors analyzed quality functions in three levels of granularity from node-level to community-level and network-level.

Guimerà et al. in [8] proposed a methodology that allows one to extract and display information about node roles in complex networks. Specifically, the role of a node in a network partition can be defined by its value of within-module connectivity and its participation into inter-cluster connections. Our work here is based on a similar method of illustration, but instead of analyzing roles of nodes in a network partition, we conduct a community-level analysis to expose the nature of communities that constitute the network.

3 Community Anatomy via Out Degree Fractions of Nodes

The idea behind quality metrics is that given a partition, they indicate how the component subgraphs fit their concepts of community. In this section, we present a methodology to analyze communities in networks based on the analysis of **Out Degree Fraction (ODF)** of their nodes. We show that communities can be classified in several structural types based on the variation of their nodes' ODFs.

3.1 Community Structures in Term of ODF

A graph $G = (V, E)$ is composed of a set of $n = |V|$ nodes and $m = |E|$ edges where $E = (u, v) : u, v \in V$. Given a cluster S of n_S nodes, which is a subgraph of G, a function $f(S)$ quantifies a quality metric of S according to a particular notion of community. Let $d(u)$ be the degree of node u. The out degree fraction of node u in community S is measured by:

$$ODF_S(u) = \frac{|(u, v) \in E : v \notin S|}{d(u)}$$

When evaluating a community, one would normally not only want to know the average fraction of out degrees in that community, but also be curious about how are they distributed over nodes. By observing the average and the standard deviation of ODF values of nodes in a community, one could deduce the composition of its population. From now on, for given a community S, meanODF and sdODF denote the average and the standard deviation of ODF values of nodes in S respectively. They are calculated as following:

- $meanODF(S) = \frac{\sum_{u \in S} ODF_S(u)}{n_S}$
- $sdODF(S) = \left(\frac{\sum_{u \in S} [ODF_S(u) - meanODF(S)]^2}{n_S - 1} \right)^{1/2}$

As a meanODF value indicates the average out degree fraction of nodes in a community, a low meanODF implies that nodes in the community connect mostly with other nodes inside the community while a high meanODF means that nodes

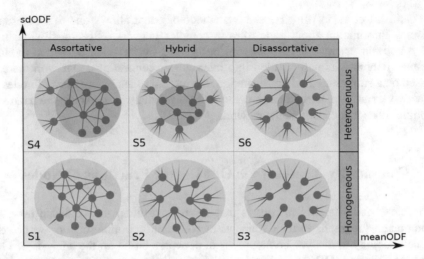

Fig. 1 Six representative community structures that can be measured by community's nodes out degree fractions (*meanODF* and *sdODF*). *Blue edges* represent intra-community connections and *red edges* (stubs) represent inter-community connections. Dark background zones in $S4, S5, S6$ structures illustrate a core-periphery arrangement

connect preferably to nodes in other communities rather than to the ones in its own. We could refer low *meanODF* and high *meanODF* characteristics to assortative structure and disassortative structure respectively. A medium value of *meanODF* in this case signifies a hybrid structure of the community as shown in Fig. 1.

We know that a standard deviation of a variable help us to understand the fluctuation of its values. Thus, to understand the composition of a community, we inspect its *sdODF* value. A low *sdODF* value implies that community's out degrees are proportionally distributed among nodes in a way to limit the variation of *ODF* values. Meanwhile, a high *sdODF* argues a diverse connection patterns of nodes in the community. In other words, based on *sdODF* value of a community, one can determine whether is there a clear division of roles [8] among nodes in the community (high *sdODF*) or nodes are just basically regular ones (low *sdODF*).

One might wonder why we chose the average and the standard deviation of *ODF* values of nodes in order to describe a community. In fact, each quality metric has its own meaning and reveals a different aspect of community structure [13]. Because the notion of community also changes according to domains of application and analysis purposes, there is actually no universal metric that can generalize the goodness of communities. Generally, one would expect a clustering where the majority of edges reside between nodes in a same cluster while there are few edges that cross to other clusters. The *meanODF* and *sdODF* are used since together, they can describe the distribution edges among nodes in an informative way. However, quality metrics could be chosen differently to match with specific concepts of community.

3.2 Community Structures Classification via Nodes' ODFs Analysis

Follow this line of argumentation, we classify communities into different structural groups based on their node orientations and their structure homogeneities. Community structures in real networks are undeniably much more complex and can not just only be described by *meanODF* and *sdODF* values. However, this simplification helps one to have a general view of networks by qualifying community anatomy. Here, we suggest to classify communities into six following groups, which are illustrated in Fig. 1:

- *Conventional communities* (S1—low *meanODF* and low *sdODF*): This structure corresponds to the traditional definition of community where the majority of edges locate inside communities. Most of actual community detection methods are based on this notion. In addition, community's out degrees are homogeneously spread over its nodes.
- *Casual communities* (S2—medium *meanODF* and low *sdODF*): Modular structure is not very clear in this type of community since there is not a clear propensity in node connections inside and outside of communities.
- *Extrovert communities* (S3—high *meanODF* and low *sdODF*): This structure exposes an explicit disassortative structure where members in a same community are not joined together generally, but rather connect with members of other communities.
- *Full-core communities* (S4—low *meanODF* and high *sdODF*): This group of communities shows a striking similarity with ones of S1 structure since both possess relatively dense inner connections. The only distinction between S1 and S4 structure is that S4 contains a few numbers of *active connector* nodes, which attract most out links. These connectors form a peripheral zone, whereas the other nodes constitute a core as illustrated in Fig. 1.
- *Half-core communities* (S5—medium *meanODF* and high *sdODF*): These communities also display core-periphery structure, but there is not anymore a quantity dominance of core nodes over periphery nodes like that of in structure S4.
- *Seed-core communities* (S6—high *meanODF* and high *sdODF*): Core-periphery structure in this class of communities is degenerated or even disappeared since out-bound connectors predominate in the whole community. Most nodes connect mainly outside their community with a few exceptions. This structure have many similarities with S3 structure and S5 structure and can be considered as a transition state of community evolution between S3 and S5.

Here, a node is considered as a core node if it connects mostly inside its community whether a periphery node is the one that attaches communities together.

3.3   Network Partition Evaluation Methodology

We propose a methodology to decompose network partitions into classes of structurally similar communities. For a given network partition:

1. Compute *meanODF* and *sdODF* values over all communities (cf. Sect. 3.1)
2. Present each community by its couple of values (*meanODF*, *sdODF*) to observe the distribution of these quality metrics.
3. Choose thresholds for each quality metric in order to describe desired structure qualities for communities.
4. Identify structure profiles of all communities based on a representative map (cf. Fig. 1) defined from step 3

As previously mentioned in Sect. 3.1, quality metrics reveal different aspects of community structures. Thus, replacing *meanODF* and *sdODF* in step 1 by other quality metrics could also provide further structural information on community structures of networks under consideration. A list of quality metrics and their performances in detecting ground-truth communities in several networks can be found in [13].

Based on requirements of each specific context, thresholds to be chosen in step 3 could be varied and must not cover the whole ranges of *meanODF* and *sdODF*. In this case, the methodology also serves as a filter to eliminate unqualified communities. The choice of thresholds is, in fact, relative and can be a reference for analysis purposes.

4   Community Description Experiment on Real-World Networks

We analyze undirected, unweighted and scale-free networks [2] with ground-truth communities on SNAP dataset [10]. These communities are overlapped and may not cover the whole network, which means one node can belong to no community or can be members of many communities at a same time. The community sizes, the overlap sizes and the community memberships per node in these networks follow a power-law distribution [13].

Livejournal network is an online blogging community where users declare their friendships. *Youtube* network represents a social network on Youtube video sharing website. *DBLP* computer science bibliography co-authorship network is constructed in a way that two authors are connected if they published at least one paper together. *Amazon* co-purchased network represents products which are frequently bought together on Amazon website. A description of these networks and measures on their ground-truth communities can be found in Table 1.

Here, we take the conductance $\bar{\mu}$ as an example to demonstrate the weaknesses of conventional quality metrics [1]. The latter represents average portion of boundary

Table 1 Network summary: N number of nodes, E number of edges, C number of communities, S average community size, O community memberships per node, $\bar{\mu}$ average conductance [13] of communities

Network	N	E	C	S	O	$\bar{\mu}$	Community nature
Livejournal[a]	4.0M	34.7M	664,414	10.79	6.24	0.95	User-defined communities
Youtube[a]	1.13M	3.0M	16,386	7.89	2.45	0.91	User-defined groups
DBLP[a]	0.32M	1.05M	13,477	53.41	2.76	0.62	Publication venues
Amazon[a]	0.33M	0.93M	75,149	30.22	7.16	0.58	Product categories

[a]http://snap.stanford.edu/data/

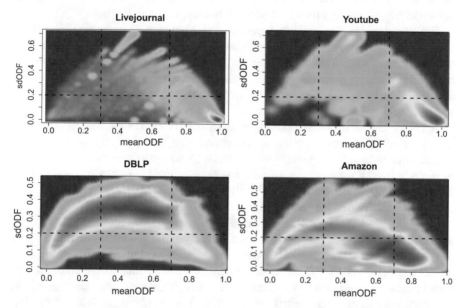

Fig. 2 The density of ground-truth communities on a *meanODF*, *sdODF* space. The *dashed lines* represent thresholds between the six presented structures *S1–S6* (cf. Sect. 3.2)

edges in ground-truth communities of a network. This metric could tell us a global score of community quality, but they can not distinguish many different structures that exist simultaneously in networks. For instance, the average conductance $\bar{\mu}$ shows that there are above 90% of edges in *Livejournal* and *Youtube* that cross communities, meanwhile these numbers are about 60% in *DBLP* and *Amazon*. However, one could not gain more insight into the differences of community structure between *Livenetwork* and *Youtube*, or between *DBLP* and *Amazon*. Thus, we describe the ground-truth communities of these networks in the next part by applying the methodology presented in Sect. 3.3.

Figure 2 presents the landscape of *meanODF*, *sdODF* values of all ground-truth communities in the four networks (cf. Sect. 3.3, step 1 and 2). We classify these communities into the six groups as presented in Sect. 3.2 by choosing thresholds for

Table 2 The composition of ground-truth communities in the four networks (in percentage). Bold values indicate dominant structure(s)

Network	S1	S2	S3	S4	S5	S6
Livejournal	0.29	0.74	**90.17**	0.31	3.88	4.61
Youtube	0.08	2.36	**65.36**	1.37	**17.55**	**13.28**
DBLP	6.28	2.07	4.87	**23.44**	**57.86**	5.48
Amazon	8.33	**31.13**	**23.57**	9.13	**26.63**	1.21

meanODF at 0.3, 0.7 and for *sdODF* at 0.2 (cf. Sect. 3.3, step 3). The landscape helps us to analyze the composition of ground-truth community structures in each network. We remind that the density landscapes in Fig. 2 do not represent the networks themselves, but the community structures in these networks.

We can see that the structural patterns of ground-truth communities within four networks are totally distinct. Normally, one would expect that ground-truth communities in a network have a quite similar structure, but the density landscapes in Fig. 2 illustrate a more complex community composition. While in *Livejournal* and *Youtube* networks, the majority of communities have a similar structure, those in *DBLP* and *Amazon* networks vary in a large range. Table 2 describes a global composition of the four networks in terms of the six basic structural groups ($S1$–$S6$). We find that $S3$ structure occupies around 90% and 65% of communities in *Livejournal* and *Youtube* networks respectively. This implies the fact that most users in these networks usually have friendships outside their communities rather than inside. In addition, there are many closely-knit members in *Youtube* network, who are not very active outside their communities ($S5$ and $S6$).

In *DBLP* and *Amazon* network, although there is always a dominance of some structures, we notice a more equilibrate repartition of communities over the landscapes. In the case of *DBLP*, nearly 60% of publication venues ($S5$) attract a variety type of authors in term of cooperation profile. These communities could represent traditional publication venues which gather at the same time high influence authors and newcomers. Meanwhile, there is about 23.44% publication venues where presented just a few active *eminent* authors. In *Amazon* network, the high presence of $S2$ and $S3$ structures explains that products are more often co-purchased with ones of other categories. Besides, there are also many miscellaneous product categories ($S1$, $S4$, $S5$) which consist of a high portion of products that are mostly complemented by ones in the same categories. Further analysis in natures and functionalities of products need to be conducted in order to understand this commercial network.

5   Conclusion

We know that optimizing a particular quality function could discard many interesting community structures. The methodology proposed in this paper presented a new approach to community analysis, where specialists can evaluate network partitions themselves according to contextual concepts with more insight into community structures. We also extended the notion of community, which is actually generalized for most community detection methods and then described communities in real networks in an informative way that many quality metrics fail to do. The extended notion could be applied in order to identify more complex structures in networks. Furthermore, we continue to enrich this notion by employing other pairs of metrics to describe more sophisticated characteristics that exist in real-world communities.

References

1. Almeida, H., Guedes, D., Meira Jr., W., Zaki, M.J.: Is there a best quality metric for graph clusters? In: Gunopulos, D., et al. (eds.) ECML PKDD 2011, Part I. LNCS, vol. 6911, pp. 44–59. Springer, Heidelberg (2011)
2. Barabási, A.L., Albert, R.: Emergence of scaling in random networks. Science **286**(5439), 509–512 (1999)
3. Clauset, A., Newman, M.E.J., Moore, C.: Finding community structure in very large networks. Phys. Rev. E **70**, 066111 (2004)
4. Creusefond, J., Largillier, T., Peyronnet, S.: On the evaluation potential of quality functions in community detection for different contexts. In: Advances in Network Science: 12th International Conference and School, NetSci-X (2016)
5. Faloutsos, M., Faloutsos, P., Faloutsos, C.: On power-law relationships of the internet topology. In: Proceedings of the Conference on Applications, Technologies, Architectures, and Protocols for Computer Communication, pp. 251–262 (1999)
6. Fire, M., Tenenboim, L., Lesser, O., Puzis, R., Rokach, L., Elovici, Y.,: Link prediction in social networks using computationally efficient topological features. In: 2011 IEEE Third International Conference on Social Computing (SocialCom), pp. 73–80 (2011)
7. Girvan, M. and Newman, M.E.J.: Community structure in social and biological networks. Proc. Natl. Acad. Sci. **99**(12), 7821–7826 (2002)
8. Guimerà, R., Amaral, L.A.N.: Cartography of complex networks: modules and universal roles. J. Stat. Mech. Theory Exp. **2005**, P02001 (2005)
9. Guimerà, R., Amaral, L.A.N.: Functional cartography of complex metabolic networks. Lett. Nat. **7028**, 895–900 (2005)
10. Leskovec, J., Krevl, A.: SNAP Datasets: Stanford Large Network Dataset Collection. http:// snap.stanford.edu/data (2014). Reference date: 13/12/2016
11. Newman, M.E., Girvan, M.: Finding and evaluating community structure in networks. Phys. Rev. E **69**(2), 026113 (2004)
12. Ravasz, E., Somera, A.L., Mongru, D.A., Oltvai, Z.N., Barabási, A.-L.: Hierarchical organization of modularity in metabolic networks. Science **297**(5586), 1551–1555 (2002)
13. Yang, J., Leskovec, J.: Defining and evaluating network communities based on ground-truth. Knowl. Inf. Syst. **42**, 181–213 (2015). http://dx.doi.org/10.1007/s10115-013-0693-z
14. Yook, S.H., Jeong, H., Barabási, A.-L.: Modeling the Internet's large-scale topology. Proc. Natl. Acad. Sci. **99**, 13382–13386 (2002)

Do Network Models Just Model Networks? On the Applicability of Network-Oriented Modeling

Jan Treur

Abstract In this paper for a Network-Oriented Modelling perspective based on temporal-causal networks it is analysed how generic and applicable it is as a general modelling approach and as a computational paradigm. This results in an answer to the question in the title different from: network models do just model networks!

Keywords Network-oriented modeling • Temporal-causal network models
• Applicability

1 Introduction

Although the notion of network itself and its use in different contexts can be traced back to the years 1930–1960 (e.g., [1], or [2], Chap. 1, Sect. 1.4), the notion of Network-Oriented Modelling as a modelling approach (sometimes also indicated by NOM) can be found only in more recent literature, and only for specific domains. More specifically, this term is used in different forms in the context of modelling organisations and social systems (e.g., [3–5]), of modelling metabolic processes (e.g., [6, 7]), and of modelling electromagnetic systems (e.g., [8]). The Network-Oriented Modelling approaches put forward in this literature are specific for the domains addressed, respectively social systems, metabolic processes and electromagnetic systems.

This and other Network Science literature may suggest that sometimes in real world domains networks occur, and by some modelling process, network models are obtained that model these given networks. That might suggest a positive answer to the question in the title: network models just model networks given in real world domains. For example, network models can be obtained for metabolic networks, brain networks, computer networks and social networks, all occurring (or conceived) in the real world. However, if networks occur in real world domains,

J. Treur (✉)
Vrije Universiteit Amsterdam, Behavioural Informatics Group, Amsterdam, The Netherlands
e-mail: j.treur@vu.nl

© Springer International Publishing AG 2017
E. Shmueli et al. (eds.), *3rd International Winter School and Conference on Network Science*, Springer Proceedings in Complexity, DOI 10.1007/978-3-319-55471-6_3

how often do they? Are networks everywhere? Or is there just a limited class of situations or processes that are conceived as networks?

This paper shows that also a different answer is possible to the question in the title. It is indicated how a generic, unified Network-Oriented Modelling method can be obtained that is applicable more generally. The Network-Oriented Modelling approach described here was developed initially with unification of modeling of human mental processes and social processes in mind. However, it has turned out that the scope of applicability has become much wider, as is shown in the current paper. Actually, it will be indicated that in this way practically all processes in the real world can be modelled from a Network-Oriented perspective, not only those processes or situations in the real world that are generally conceived as networks. This provides a negative answer on the question in the title: network models can model all kinds of processes in the real world, not just processes generally conceived as networks.

The Network-Oriented Modelling approach considered here uses temporal-causal networks as a basis [2, 9]. The temporal perspective allows to model the dynamics of the interaction processes within networks, and also of network evolution. Temporal-causal network models can be represented in two equivalent manners: by a conceptual representation, or by a numerical representation. Conceptual representations can have a graphical form (as a labeled graph with states as nodes and connections as arcs), or the form of a matrix. The following three elements define temporal-causal networks, and are part of a conceptual representation of a temporal-causal network model:

- *Connection weight* $\omega_{X,Y}$ Each connection from a state X to a state Y has a *connection weight* $\omega_{X,Y}$ representing the strength of the connection, often between 0 and 1, but sometimes also below 0 (negative effect).
- *Combination function* $c_Y(.)$ For each state Y (a reference to) a *combination function* $c_Y(.)$ is chosen to combine the causal impacts of other states on state Y. This can be a standard function from a library (e.g., a scaled sum function) or an own-defined function.
- *Speed factor* η_Y For each state Y a *speed factor* η_Y is used to represent how fast a state is changing upon causal impact. This is usually assumed to be in the [0, 1] interval.

Combination functions in general are similar to the functions used in a static manner in the (deterministic) Structural Causal Model perspective described, for example, in [10–12], but in the Network-Oriented Modelling approach described here they are used in a dynamic manner, as will be pointed out below briefly. Combination functions can have different forms. The more general issue of how to combine multiple impacts or multiple sources of knowledge occurs in various forms in different areas, such as the areas addressing imperfect reasoning or reasoning with uncertainty or vagueness. For example, in a probabilistic setting, for modelling multiple causal impacts on a state often independence of these impacts is assumed, and a product rule is used for the combined effect; e.g., [6]. In the areas addressing modelling of uncertainty also other combination rules are used, for example, in possibilistic approaches minimum- or maximum-based combination rules are used; e.g., [6, 13, 14]. In another area, addressing modelling based on neural networks yet another way of combining effects is used often. In that area, for combination of the impacts of multiple neurons on a given neuron usually a logistic sum function is used; e.g., [15–17].

The applicability of a specific combination rule may depend much on the type of application addressed, and even on the type of states within an application. Therefore the Network-Oriented Modelling approach based on temporal-causal networks incorporates for each state, as a kind of label or parameter, a combination function indicating a way to specify how multiple causal impacts on this state are aggregated. For this aggregation a number of standard combination functions are available as options; for more details, see [2], Chap. 2, Sects. 2.6 and 2.7). These options cover, for example, scaled sum functions, logistic sum functions, product functions and max and min functions. In addition, there is still the option to specify any other (non-standard) combination function.

A conceptual representation of temporal-causal network model, including the above three concepts (connection weight, combination function, speed factor) can be transformed in a systematic and automated manner into an equivalent numerical representation of the model [2, 9] by composing the following *difference* and *differential equation* for each state Y (where the X_i are the states with connections to Y):

$$Y(t + \Delta t) = Y(t) + \eta_Y \left[\mathbf{c}_Y \left(\omega_{X_1}, {}_YX_1(t), \dots, \omega_{X_k}, {}_YX_k(t) \right) - Y(t) \right] \Delta t$$

$$dY(t)dt = \eta_Y \left[\mathbf{c}_Y \left(\omega_{X_1}, {}_YX_1(t), \dots, \omega_{X_k}, {}_YX_k(t) \right) - Y(t) \right]$$

This paper discusses how generic and applicable Network-Oriented Modeling based on temporal-causal networks is in general as a dynamic modelling approach both for continuous systems (Sect. 2) and discrete systems (Sect. 3). In Sect. 4 a number of actual applications of Network-Oriented Modeling is discussed, varying from mental processes to social interaction processes.

2 Modeling Continuous Dynamical Systems as Networks

In the current section it is discussed how temporal-causal networks subsume smooth continuous dynamical systems, as advocated, for example in [18] to model human mental processes. The notion of *state-determined system*, adopted from [19] was taken as the basis to describe what a dynamical system is in [20], p. 6. That a system is state-determined means that its *current state always determines a unique future behaviour*. This property is reflected in modelling and simulation, as usually some *rules of evolution* are specified and applied that indicate how exactly a future state depends on the current state. State-determined systems can be specified in mathematical formats; see [19], pp. 241–252 for some details. A finite set of states (or variables) $X_1, \dots, X_n$ is assumed describing the system's dynamics via functions $X_1(t), \dots, X_n(t)$ of time variable t.

In this section it is shown how any smooth continuous dynamical system (assumed as state-determined) can be modeled by a temporal-causal network in two steps. First it is discussed how any smooth continuous state-determined system can be described by a set of first-order differential equations, and next it is shown how any set of first-order differential equations can be modelled as a temporal-causal network.

2.1 From State-Determined Systems to Differential Equations

From an abstract theoretical perspective the state-determined system criterion can be formalized in a numerical manner by functions $F_i(X_1, \ldots, X_n, s)$ that express how for each time point t the future value of each state X_i at time $t + s$ uniquely depends on s and on $X_1(t), \ldots, X_n(t)$; see also [2], Chap. 2, Sect. 2.9; for an alternative treatment, see [19], pp. 243–244. To illustrate the idea by a simple example, consider a state-determined system in one state variable X with values ≥ 0 described by

$$X(t + s) = F(X(t), s) = \sqrt{\left(X(t)^2 + \alpha s\right)}$$

By differentiating both sides to s and by choosing $s = 0$ the following is obtained:

$$\frac{dX(t)}{dt} = \left[\frac{\partial F(X(t), s)}{\partial s}\right]_{s=0}$$

The right hand side can be worked out as follows:

$$\frac{\partial F(X(t), s)}{\partial s} = \frac{\partial \sqrt{\left(X(t)^2 + \alpha s\right)}}{\partial s} = \frac{\frac{1}{2}\alpha}{\sqrt{\left(X(t)^2 + \alpha s\right)}}$$

So by substituting $s = 0$:

$$\left[\frac{\partial F(x(t), s)}{\partial s}\right]_{s=0} = \left[\frac{\frac{1}{2}\alpha}{\sqrt{X(t)^2}}\right] = \frac{\frac{1}{2}\alpha}{X(t)}$$

Thus the following differential equation for X is obtained:

$$\frac{dX(t)}{dt} = \frac{\frac{1}{2}\alpha}{X(t)}$$

This differential equation has an analytic solution of the form

$$X(s) = \sqrt{\left(X(0)^2 + \alpha s\right)}$$

which indeed confirms the formula assumed at the start of the example. This illustrates how a state-determined system can be described by first-order differential equations. For more details, see [2], Chap. 2, Sect. 2.9; for an alternative treatment, see [19], pp. 243–244.

2.2 From Differential Equations to Temporal-Causal Networks

Next it is shown by an example how any model described by a set of first-order differential equations can be described by a temporal-causal network. Consider an arbitrary example of a model described by a set of first-order differential equations:

$$\frac{dW(t)}{dt} = Z(t) - Y(t)\,(1 - W(t))$$

$$\frac{dX(t)}{dt} = X(t)\,(1 - W(t))$$

$$\frac{dY(t)}{dt} = X(t) - Y(t) + Z(t)$$

$$\frac{dZ(t)}{dt} = Z(t)\,(1 - Y(t))$$

To determine a temporal-causal network representation for this model, the four states W, X, Y, Z are considered as the nodes. From each of the equations by inspecting which states occur in the right hand side it can subsequently be determined that (in addition to the effect of the state itself): Y and Z affect W, W affects X; X and Z affect Y; Y affects Z. These causal connections can be represented in a conceptual graphical form that is shown in Fig. 1. Note that the connection weights and speed factors are not mentioned as they are all assumed 1. The combination functions will be discussed below. Considering the numerical representation, note that, when comparing, for example, the second differential

Fig. 1 Graphical conceptual representation for the example model based on the given differential equation representation

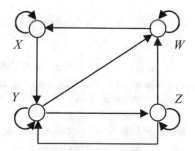

equation to the numerical representation format defined in Sect. 1, it can be rewritten as.

$$\frac{dX(t)}{dt} = X(t)\,(1 - W(t)) = [X(t)\,(1 - W(t)) + X(t)] - X(t)$$

Here $[X(t)\,(1 - W(t)) + X(t)]$ can be viewed as the result of a combination function

$$c_X\,(V_1, V_2) = V_2\,(1 - V_1) + V_2$$

applied to $X(t)$ (for V_1) and $W(t)$ (for V_2). In a similar manner the following combination functions can be identified from the differential equations:

$$c_W\,(V_1, V_2, V_3) = V_3 - V_2\,(1 - V_1) + V_1 = V_1 - V_2 + V_3 + V_1 V_2$$

$$c_X\,(V_1, V_2) = V_2\,(1 - V_1) + V_2 = 2V_2 - V_1 V_2$$

$$c_Y\,(V_1, V_2, V_3) = V_1 - V_2 + V_3 + V_2 = V_1 + V_3$$

$$c_Z\,(V_1, V_2) = V_2\,(1 - V_1) + V_2 = 2V_2 - V_1 V_2$$

Using these combination functions, the original differential equations transform into the following numerical representation of a temporal-causal network representation where all speed factors η and all connection weights ω for connected states are 1:

$$\frac{dW(t)}{dt} = \eta_W\,[c_W\,(\omega_{Y,W}Y(t), \omega_{Z,W}Z(t), \omega_{W,W}W(t)) - W(t)]$$

$$\frac{dX(t)}{dt} = \eta_X\,[c_X\,(\omega_{X,X}X(t), \omega_{W,X}W(t)) - X(t)]$$

$$\frac{dY(t)}{dt} = \eta_Y\,[c_Y\,(\omega_{X,Y}X(t), \omega_{Y,Y}Y(t), \omega_{Z,Y}Z(t)) - Y(t)]$$

$$\frac{dZ(t)}{dt} = \eta_Z\,[c_Z\,(\omega_{Y,Z}Y(t), \omega_{Z,Z}Z(t)) - Z(t)]$$

It turns out that the model described by the differential equations can be remodeled as a special case of a more general numerical temporal-causal network model representation.

So, it has been found that any smooth continuous dynamical system can be modelled as a temporal-causal network model, by choosing suitable parameters such as connection weights, speed factors and combination functions. In this sense this Network-Oriented Modelling approach is as generic as dynamic modelling approaches put forward, for example, in [15–19, 21]. This indicates that using this Network-Oriented Modelling approach does not limit the scope of applicability of the modelling in comparison to the general (smooth continuous) dynamical system approach. In Sect. 3 the discrete case is analysed.

3 Modeling Discrete Dynamical Systems as Networks

The numerical representations of temporal-causal network models can also be used to model any discrete and binary processes, as will be shown in this section.

3.1 Real-Valued Discrete Dynamical Systems

To consider discrete dynamical systems as often considered in discrete event simulation (e.g., [22, 23]), for example, first set time step $\Delta t = 1$. Then the difference equation for any state Y becomes

$$Y(t+1) = Y(t) + \eta_Y \left[\mathbf{c}_Y \left(\omega_{X_1, Y} X_1(t), \ldots, \omega_{X_k, Y} X_k(t) \right) - Y(t) \right]$$
$$= (1 - \eta_Y) Y(t) + \eta_Y \mathbf{c}_Y \left(\omega_{X_1, Y} X_1(t), \ldots, \omega_{X_k, Y} X_k(t) \right)$$

As $0 \leq \eta_Y \leq 1$ is assumed here, the new value for Y is a weighted average of the current value and the aggregated impact with η_Y and $(1 - \eta_Y)$ as weights. Next, if the connection weights for all states X and Y with a connection from X to Y is assumed $\omega_{X,Y} = 1$, the following is obtained:

$$Y(t+1) = (1 - \eta_Y) Y(t) + \eta_Y \mathbf{c}_Y \left(X_1(t), \ldots, X_k(t) \right)$$

Moreover, if $\eta_Y = 1$ for all states Y is assumed, the following is obtained:

$$Y(t+1) = \mathbf{c}_Y \left(X_1(t), \ldots, X_k(t) \right)$$

This is a very general format, often used to specify iteration rules for discrete simulations. So, all such approaches are covered by temporal-causal networks.

3.2 Binary Discrete Dynamical Systems and Finite State Machines

One step further is when all state values are assumed binary: 0 or 1, and all combination functions $c_Y(\ldots)$ only generate values 0 or 1, when applied to values 0 or 1. Then the previous iteration equation

$$Y(t+1) = c_Y(X_1(t), \ldots, X_k(t))$$

can be taken as a general evolution or transition rule for a discrete binary dynamical system. If the overall states are defined as vectors $X(t) = (X_1(t), \ldots, X_k(t))$ with values 0 or 1, and for $V = (V_1, \ldots, V_k)$ the vector combination function $c(.)$ is defined by

$$c(V) = (c_{X1}(V), \ldots, c_{Xk}(V)) = (c_{X_1}(X_1(t), \ldots, X_k(t)), \ldots, c_{X_k}(X_1(t), \ldots, X_k(t))),$$

the transitions of overall states are defined as

$$(X_1(t+1), \ldots, X_k(t+1)) = (c_{X_1}(X_1(t), \ldots, X_k(t)), \ldots, c_{X_k}(X_1(t), \ldots, X_k(t)))$$

or in short notation

$$X(t+1) = c(X(t))$$

This is illustrated by a simple model of traffic lights at a crossing of two roads A and B, where traffic on A has priority over traffic on B. For example, if no approaching traffic is sensed on road A, then the traffic light for road B is not red, and for road A red. The rules describing state transitions can be described by the following transition relations:

> traffic_on_road_A → no red_light_for_road_A ∧ red_light_for_road_B
> no traffic_on_road_A ∧ traffic_on_road_B → no red_light_for_road_B ∧
>   red_ light_for_road_A
> no traffic_on_road_A ∧ no traffic_on_road_B → no red_light_for_road_A ∧
>   red_light_for_road_B

These transition relations can be represented by a (vector) combination function defined by: $c(1, V_2, V_3, V_4) = (1, V_2, 0, 1)$; $c(0, 1, V_3, V_4) = (0, 1, 1, 0)$; $c(0, 0, V_3, V_4) = (0, 0, 0, 1)$. This shows how the Network-Oriented Modelling approach based on temporal-causal networks subsumes modelling by discrete binary dynamical systems.

Within theoretical analyses often variants of transition systems or finite state machines are used as universal ways to specify computational processes. In more detail and illustrated by the above traffic light example, the format for binary discrete dynamical systems described above as a special case of temporal-causal networks can be used to model transition systems or finite state machines in the format of this Network-Oriented Modeling approach. This can be done by assuming that states are described by vectors X based on a number of binary state variables X_i (with values 0 or 1) and by defining $c(X) = X'$ if and only if within a given finite state machine or transition system there is a transition from the overall state represented by vector X to the overall state represented by vector X'. As finite state machines and transition systems are often considered to be general computational formats, this shows how very wide classes of computational processes can be covered by Network-Oriented Modeling based on temporal-causal networks.

4   Some Actual Applications of Network-Oriented Modeling

In [2] in a number of chapters applications of Network-Oriented Modeling based on temporal-causal networks for the area of human mental and social processes are discussed. In Part II in Chapters 3 to 6 models are discussed that address the way in which emotions are integrated in an interactive manner in practically all mental processes. In this it is discussed how within Cognitive, Affective and Social Neuroscience mechanisms have been found that indicate how emotions interact in a bidirectional manner with many other mental processes and behaviour. Based on this, an overview of neurologically inspired temporal-causal network models for the dynamics and interaction for emotions is discussed. Thus an integrative perspective is obtained that can be used to describe, for example, how emotions interact with beliefs, experiences, decision making, and emotions of others, and also how emotions can be regulated. It is pointed out how integrated temporal-causal network models of such mental processes incorporating emotions can be obtained.

In Chap. 4 it is discussed how emotions play a role in generating dream episodes from a perspective of internal simulation. Building blocks for this internal simulation are memory elements in the form of sensory representations and their associated emotions. In the presented temporal-causal network model, under influence of associated feeling levels and mutual competition, some sensory representation states pop up in different dream episodes. As a form of emotion regulation the activation levels of both the feelings and the sensory representation states are suppressed by control states. In Chap. 5 it is discussed how dreaming is used to learn fear extinction. Fear extinction has been found not to involve weakening of fear associations, but instead it involves the strengthening of fear suppressing connections that form a counter balance against the still persisting fear associations. To this end neural mechanisms are used that strengthen these suppressing connections, as a form of learning of emotion regulation. The presented adaptive temporal-causal network model based on Hebbian learning addresses this adaptation process.

Chapter 6 addresses the role of emotions in rational decision making. It has been found that neurological mechanisms involving emotions play an important role in rational decision making. In this chapter an adaptive temporal-causal network model for decision making based on predictive loops through feeling states is presented, where the feeling states function in a process of valuing of decision options. Hebbian learning is considered for different types of connections in the adaptive model. Moreover, the adaptive temporal-causal network model is analysed from the perspective of rationality. To assess the extent of rationality, measures are introduced reflecting what would be rational for a given environment's characteristics and behaviour. It is shown how during the adaptive process this model for decision making achieves higher levels of rationality.

Part III of [2], consisting of Chap. 7–11, focuses on persons functioning in a social context. In Chap. 7 an overview is presented of a number of recent findings from Social Neuroscience on how persons can behave in a social manner. For example, shared understanding and collective power are social phenomena that serve as a form of glue between individual persons. They easily emerge and often involve both cognitive and affective aspects. As the behaviour of each person is based on complex internal mental processes involving, for example, own goals, emotions and beliefs, it would be expected that such forms of sharedness and collectiveness are very hard to achieve. From a neurological perspective, mirror neurons and internal simulation are core concepts to explain the mechanisms underlying such social phenomena. In this chapter it is discussed how based on such neurological concepts temporal-causal network models for social processes can be obtained. It is discussed how these models indeed are an adequate basis to simulate the emergence of shared understanding and collective power in groups.

Within a social context the notion of ownership of actions is important. Chapter 8 addresses this notion. It is related to mechanisms underlying self-other distinction, where a self-ownership state is an indication for the self-relatedness of an action and an other-ownership state to an action attributed to someone else. The temporal-causal network model presented in this chapter generates prior and retrospective ownership states for an action based on principles from recent neurological theories. A prior self-ownership state is affected by prediction of the effects of a prepared action as a form of internal simulation, and exerts control by strengthening or suppressing actual execution of the action. A prior other-ownership state also plays a role in mirroring and analysis of an observed action performed by another person, without imitating the action. A retrospective self-ownership state depends on whether the sensed consequences of an executed action co-occur with the predicted consequences, and is the basis for acknowledging authorship of actions in social context. Scenarios are shown for vetoing a prepared action due to unsatisfactory predicted effects. Moreover, it is shown how poor action effect prediction capabilities can lead to reduced retrospective ownership states, for example, in persons suffering from schizophrenia. This can explain why sometimes the own actions are attributed to others, or actions of others are attributed to oneself.

Chapter 9 addresses how in social interaction between two persons usually each person shows empathic understanding of the other person. This involves both nonverbal and verbal elements, such as bodily expressing a similar emotion and verbally expressing beliefs about the other person. Such social interaction relates to an underlying neural mechanism based on a mirror neuron system and self-other distinction. Differences in social responses of individuals can often be related to differences in functioning of certain neurological mechanisms, as can be seen, for example, in persons with a specific type of Autism Spectrum Disorder (ASD). This chapter presents a temporal-causal network model which, depending on personal characteristics, is capable of showing different types of social response patterns based on such mechanisms, adopted from theories on the role of mirror neuron systems, emotion integration, emotion regulation, and empathy in ASD. The personal characteristics may also show variations over time. This chapter also addresses this adaptation over time. To this end it includes an adaptive temporal-causal network model capable of learning social responses, based on insights from Social Neuroscience.

Chapter 10 addresses joint decision making. The notion of joint decision making as considered does not only concern a choice for a common decision option, but also sharing a good feeling and mutually acknowledged empathic understanding about it. The model is based on principles from recent neurological theories on mirror neurons, internal simulation, and emotion-related valuing. Emotion-related valuing of decision options and mutual contagion of intentions and emotions between persons are used as a basis for mutual empathic understanding and convergence of decisions and their associated emotions.

In Chap. 11 it is discussed how adaptive temporal-causal network models can be used to model evolving social interactions. This perspective simplifies persons to just one state and expresses the complexity in the structure of the social interactions, modelled by a network. The states can represent, for example, a person's emotion, a belief, an opinion, or a behaviour. Two types of dynamics are addressed: dynamics based on a fixed structure of interactions (modelled by a non-adaptive temporal-causal network model), and dynamics where the social interactions themselves change over time (modelled by an adaptive temporal-causal network model). In the case of an adaptive network model, the network connections change, for example their weights may increase or decrease, or connections are added or removed. Different types of adaptive social network models are addressed, based on different principles: the homophily principle assuming that connections strengthen more when the persons are more similar in their state (the more you are alike, the more you like each other), and the more becomes more principle assuming that persons that already have more and stronger connections also attract more and stronger connections.

5 Discussion

The Network-Oriented Modelling approach based on temporal-causal networks as discussed here, provides a modelling approach that enables a modeller to design high level conceptual model representations in the form of (cyclic) labelled graphs, which can be systematically transformed in an automated manner into numerical representations that can be used to perform simulation experiments.

It sometimes is a silent assumption that a Network-Oriented Modeling approach can only work for specific application domains, where networks are more or less already given or conceived in the real world. This paper shows that this not exactly a correct assumption. It has been shown that the applicability of the Network-Oriented Modeling approach based on temporal-causal networks is very wide; for example, it subsumes modelling approaches based on the dynamical system perspective [18, 20] often used to obtain cognitive models, and modelling approaches based on discrete (event) and agent simulation [22, 23]. This provides a different light on the question in the title of this paper, different from network models do just model networks!

References

1. Treur, J.: Network-oriented modelling and its conceptual foundations. In: Proceeding of the 8th International Conference on Social Informatics, SocInfo'16. Lecture Notes in AI. Springer Publishers, Barcelona (2016)
2. Treur, J.: Network-Oriented Modeling: Addressing Complexity of Cognitive, Affective and Social Interactions. Understanding Complex Systems Series. Springer, Barcelona (2016)
3. Chung, B., Choi, H., Kim, S.: Workflow-enabled internet service delivery for a variety of access networks. In: The 7th Asia-Pacific Network Operations and Management Symposium, APNOMS (2003)
4. Naudé, A., Le Maitre, D., de Jong, T., Mans, G. F. G., & Hugo, W.: Modelling of spatially complex human-ecosystem, rural-urban and rich-poor interactions (2008)
5. Elzas, M.S.: Organizational structures for facilitating process innovation. In: Real Time Control of Large Scale Systems, pp. 151–163). Springer, Berlin (1985)
6. Dubois, D., Lang, J., Prade, H.: Fuzzy sets in approximate reasoning, part 2: logical approaches. Fuzzy Sets Syst. **40**, 203–244 (1991)
7. Cottret, L., Jourdan, F.: Graph methods for the investigation of metabolic networks in parasitology. Parasitology. **137**, 1393–1407 (2010)
8. Russer, P., Cangellaris, A.C.: Network-oriented modeling, complexity reduction and system identification techniques for electromagnetic systems. In Proceeding of the 4th International Workshop on Computational Electromagnetics in the Time-Domain: TLM/FDTD and Related Techniques, pp. 105–122 (2001)
9. Treur, J.: Dynamic modelling based on a temporal-causal network modelling approach. Biol. Inspired Cognit. Archit. **16**, 131–168 (2016)
10. Mooij, J.M., Janzing, D., Schölkopf, B.: From differential equations to structural causal models: the deterministic case. In: Nicholson, A., Smyth, P. (eds.) Proceedings of the 29th Annual Conference on Uncertainty in Artificial Intelligence (UAI-13), pp. 440–448. AUAI Press, Corvallis (2011)
11. Pearl, J.: Causality. Cambridge University Press, New York (2000)
12. Wright, S.: Correlation and Causation. J. Agric. Res. **20**, 557–585 (1921)
13. Dubois, D., Prade, H.: Possibility theory, probability theory and multiple-valued logics: a clarification. Ann. Math. Artif. Intell. **32**, 35–66 (2002)
14. Zadeh, L.: Fuzzy sets as the basis for a theory of possibility. Fuzzy Sets and Systems, **1**, 3–28, (1978). (Reprinted in Fuzzy Sets and Systems 100 (Supplement): 9–34, 1999)
15. Grossberg, S.: On learning and energy-entropy dependence in recurrent and nonrecurrent signed networks. J. Stat. Phys. **1**, 319–350 (1969)
16. Hirsch, M.: Convergent activation dynamics in continuous-time networks. Neural Netw. **2**, 331–349 (1989)

17. Hopfield, J.J.: Neurons with graded response have collective computational properties like those of two-state neurons. Proc. Nat. Acad. Sci. U.S.A. **81**, 3088–3092 (1984)
18. Port, R.F., van Gelder, T.: Mind as Motion: Explorations in the Dynamics of Cognition. MIT Press, Cambridge (1995)
19. Ashby, W.R.: Design for a Brain. Chapman and Hall, London (second extended edition). First edition, 1952 (1960)
20. van Gelder T., Port, R.F.: It's about time: An overview of the dynamical approach to cognition. In: Port, R.F., van Gelder, T.: Mind as Motion: Explorations in the Dynamics of Cognition, pp. 1–43. MIT Press, Cambridge (1995)
21. Funahashi, K., Nakamura, Y.: Approximation of dynamical systems by continuous time recurrent neural networks. Neural Netw. **6**, 801–806 (1993)
22. Sarjoughian, H., Cellier, F.E. (eds.): Discrete Event Modeling and Simulation Technologies: A Tapestry of Systems and AI-Based Theories and Methodologies. Spring-Verlag, Berlin (2001)
23. Uhrmacher, A., Schattenberg, B.: Agents in discrete event simulation. In: Proceedings of the European Symposium on Simulation (ESS '98, Nottingham, England, Oct.). Society for Computer Simulation, San Diego (1998)

Visibility of Nodes in Network Growth Models

Siddharth Pal, Soham De, Tanmoy Chakraborty, and Ralucca Gera

Abstract Many real-world complex networks can be synthesized using *growth models*, where nodes enter the network at discrete time steps and attach with existing nodes based on their degree, or fitness, or a combination of both. While, the literature has mostly focused on the asymptotic global properties of such models, e.g., degree distribution, we intend to drive the focus towards investigating the dynamics from the perspective of individual nodes. In this paper, we study how the *visibility* of a node, i.e., the probability of the node to form new connections, changes over time. In particular, we study three well-known network growth models—preferential attachment, additive and multiplicative fitness models, and focus primarily on "influential nodes" or "leaders" to understand how their visibility changes over time. We present a thorough analytical study and validate our claims through simulations. Our primary finding is that influential nodes in multiplicative growth models can attain and maintain high visibility over time compared to other models; something that might not be apparent by simply looking at global network properties or other local node-centric properties.

Keywords Complex networks • Preferential attachment • Fitness models • Node visibility

This document does not contain technology or technical data controlled under either the U.S. International Traffic in Arms Regulations or the U.S. Export Administration Regulations.

S. Pal (✉)
Raytheon BBN Technologies, Cambridge, MA 02138, USA
e-mail: siddharth.pal@raytheon.com

S. De • T. Chakraborty
University of Maryland, College Park, MD 20742, USA
e-mail: sohamde@cs.umd.edu; tanchak@umiacs.umd.edu

R. Gera
Naval Postgraduate School, Monterey, CA 93943, USA
e-mail: rgera@nps.edu

© Springer International Publishing AG 2017
E. Shmueli et al. (eds.), *3rd International Winter School and Conference on Network Science*, Springer Proceedings in Complexity, DOI 10.1007/978-3-319-55471-6_4

1   Introduction

Complex networks [8, 9] are often used to describe structure and emergence of many real-world systems such as social, information, technological and biological systems. The analysis of leaders and their behavior in these networks can yield valuable insights into the understanding of how influential entities in real-world networks attract and maintain significant presence over time. For instance, influential papers in citation networks continue to acquire new citations every year, and likewise, celebrities in online social networks keep increasing their follower count over time. In order to conduct such network analysis, we study the temporal behavior of nodes in a network. We introduce a notion of *visibility of a node*, defined as the probability of that node to form new connections in a growing network. For instance, in a preferential attachment model [1, 2], the visibility of a node is proportional to its degree, and inversely proportional to the number of edges in the network. An essential aspect of the study is to investigate the *visibility profile* of a node which characterizes the temporal evolution of the node's visibility as the network grows. We argue that studying the visibility profile of nodes leads to a better understanding of network evolution due to attachment dynamics, which might not be possible to obtain by simply analysing global network properties such as degree distribution or local node-centric properties such as degree, clustering coefficient, etc. Much like node persistence over time studied in [10], our approach allows to make headway into this understanding by characterizing the *visibility behavior of leaders* in that network. While, the framework is applicable to arbitrary nodes as well, it is more interesting to first understand the leaders' behavior.

We study the visibility of high degree nodes in the Barabási-Albert (BA) model, *aka* preferential model, which explains power law behavior in real-world networks through the "rich get richer" [2] phenomenon. A few years after the introduction of the BA model, Bianconi and Barabási [3] proposed a new class of growth models whose attachment mechanism was driven by inherent properties of nodes such as novelty, usefulness, etc., captured through a fitness value. This was inspired by the "fit get richer" phenomenon observed in real-world network [3]. Subsequently, Ergun and Rodgers [4] analysed the degree distribution of the resultant growth models, when the attachment mechanism combines the degree and fitness information in an additive and multiplicative fashion. Here, we compare and contrast the visibility behaviors of the leaders in the additive and multiplicative fitness models with that in the BA model.

In particular, we seek to address the following question: Given an influential node with a high visibility at a certain point in time, how would its visibility evolve across different network growth models? One of our primary theoretical findings (mentioned in Sect. 3) is that, multiplicative fitness model allows more visibility to leaders in comparison to the additive fitness and the BA models. Furthermore, in an expected sense, influential nodes can improve their visibility over time, as long as their inherent fitness remains large in comparison to the present network; whereas, for the other two network models, the visibility is shown to always decrease in an expected sense. Simulation results (presented in Sect. 4) also support the theoretical analysis that multiplicative fitness models are better suited for explaining continued influence of leaders in certain networks.

2 Preliminaries

Consider the following sequence of graphs $\{\mathbb{G}_t, \ t = 0, 1, \ldots\}$, where $\mathbb{G}_t = (V_t, \mathbb{E}_t)$, with V_t and $\mathbb{E}_t$ being the set of nodes and edges in $\mathbb{G}_t$ respectively. In a network growth model, we have $V_t \subset V_{t+1}$ and $\mathbb{E}_t \subseteq \mathbb{E}_{t+1}$ for every $t = 0, 1, \ldots$. In other words, new nodes arrive at every time step t, and form connections with existing nodes, thus adding to the edge set of the previous graph $\mathbb{G}_{t-1}$. For purposes of simplicity, here we consider the basic model where a single node enters at any time step t, and forms a connection with one node in the existing graph $\mathbb{G}_{t-1}$. Therefore, we can label the incoming node by the time index of its entry to the network, which leads to $V_t = \{0, 1, \ldots, t\}$ for $t = 0, 1, \ldots$. Note that all our results can be easily extended to more general scenarios where multiple nodes can enter the network and incoming nodes can form multiple connections. At time t, let the degree of the node i in V_t be denoted by $D_t(i)$. Also, let the rv S_{t+1} denote the node with which an incoming node $t + 1$ connects.

Barabási-Albert (BA) Model In the preferential attachment mechanism [2], new nodes connect preferentially to existing nodes with higher degree. Let $\mathbf{p}^{BA}(t+1) = (p_i^{BA}(t+1), \ i \in V_t)$ be the pmf with which the new node indexed as $t + 1$ connects with the existing graph $\mathbb{G}_t$, i.e., $p_i^{BA}(t+1)$ is the probability with which node $t + 1$ connects with an existing node i. This is given by:

$$p_i^{BA}(t+1) = \mathbf{P}[S_{t+1} = i \mid \mathscr{F}_t] = \frac{D_t(i)}{\sum_{j \in V_t} D_t(j)}, \ i \in V_t. \tag{1}$$

where $\mathscr{F}_t$ is the σ-field generated by all the relevant random variables till time t. Note that we term node i's *visibility* in the graph $\mathbb{G}_t$ by $p_i^{BA}(t+1)$.

Fitness Based Attachment Rules In fitness based models [3, 4, 7], every node is assumed to have a fitness value independently drawn from a distribution. Assume a sequence of i.i.d. fitness rvs $(\xi, \xi_t, \ t = 0, 1, \ldots)$. In the additive fitness attachment rules, it is assumed that new nodes connect preferentially with existing nodes having a higher sum of degree and fitness value. For $t = 0, 1, \ldots$, let the pmf delineating formation of new connections at time $t + 1$ be given by $\mathbf{p}^{AF}(t+1)$, where

$$p_i^{AF}(t+1) = \frac{\xi_i + D_t(i)}{\sum_{j \in V_t}(\xi_j + D_t(j))}, \ i \in V_t. \tag{2}$$

Similarly the attachment rule for multiplicative fitness model is given by

$$p_i^{MF}(t+1) = \frac{\xi_i \cdot D_t(i)}{\sum_{j \in V_t} \xi_j \cdot D_t(j)}, \ i \in V_t. \tag{3}$$

Therefore, the *visibilities* of node i in graph $\mathbb{G}_t$ are given by $p_i^{AF}(t+1)$ and $p_i^{MF}(t+1)$ for the additive and multiplicative fitness models respectively. Note that the influential nodes as described in Sect. 1 relates to nodes having high visibility as defined for the particular network growth model in question.

3 Analytical Results on Node Visibility

In this section we first compare how the visibility of a particular node varies across the three models for a given degree sequence. Subsequently, we study and compare the evolution of node visibility over time for each of the three models.

3.1 Comparison of Node Visibility Across Growth Models

We state a lemma which compares the node visibilities across the three growth models for any given graph. While we note that the different growth models will lead to any specific graph with varying probabilities, we still conduct this exercise to obtain some intuition as to how the node visibilities compare across growth models.

Lemma 1 *For every $t = 0, 1, \ldots$, and i in V_t: Let $\mathbb{G}_t$ be the graph at time t, we have*

$$\text{(i)} \qquad p_i^{MF}(t+1) > p_i^{BA}(t+1) \ if \ \xi_i > \frac{\sum_{j \in V_t} \xi_j D_t(j)}{\sum_{\ell \in V_t} D_t(j)} \tag{4}$$

$$\text{(ii)} \qquad p_i^{AF}(t+1) > p_i^{BA}(t+1) \ if \ \xi_i > D_t(i) \frac{\sum_{j \in V_t} \xi_j}{2t} \tag{5}$$

$$\text{(iii)} \qquad p_i^{MF}(t+1) > p_i^{AF}(t+1) \ if$$

$$\xi_i \left[\sum_{j \in V_t} \xi_j \left(D_t(i) - D_t(j) \right) \right] + D_t(i) \sum_{j \in V_t} \left(\xi_i D_t(i) - \xi_j D_t(j) \right) > 0 \tag{6}$$

Proof We do the pairwise comparison of node visibilities of the three models using their expression defined in (1)–(3). Simple algebraic manipulations yields the result.

From (4), it is evident that nodes with high fitness will have greater visibility in the multiplicative fitness model than the BA model. On the other hand, from (5), it is apparent that a particular node in the additive fitness model can have higher visibility compared to the preferential attachment setting if it has a significantly high fitness value. Observe that as t increases, if $\mathbb{E}[\xi] < \infty$, the condition (5) could be approximated as $\xi_i > D_t(i)\frac{\mathbb{E}[\xi]}{2}$, which invariably gets violated as the degree of node i increases beyond the threshold $\frac{2\xi_i}{\mathbb{E}[\xi]}$. Therefore, an interesting conclusion is that in the long run, when compared with additive fitness model, the BA model will have greater visibility for influential nodes. Furthermore, if node i has high fitness value ξ_i and degree $D_t(i)$ compared to the rest of the nodes in $\mathbb{G}_t$, then condition (6) will be satisfied, leading to greater visibility of the node in the multiplicative fitness model compared to the additive fitness model.

3.2 Change in Node Visibility Over Time

The following lemma describes the change in visibility with time for the three growth models. First, we introduce some notation: Define $\Xi_t = \sum_{i \in V_t} \xi_i$ and $\chi_t = \sum_{i \in V_t} \xi_i D_t(i)$, for $t = 0, 1, \ldots$.

Lemma 2 *Let $\mathbb{G}_{t-1}$ be the graph at time $t-1$. For every $t = 0, 1, \ldots$, and i in V_{t-1}:*

(i) $$p_i^{BA}(t+1) - p_i^{BA}(t) = \frac{(t-1)\mathbf{1}[S_t = i] - D_{t-1}(i)}{2t(t-1)} \tag{7}$$

and $$\mathbb{E}\left[p_i^{BA}(t+1) - p_i^{BA}(t) \mid \mathscr{F}_{t-1}\right] = -\frac{D_{t-1}(i)}{4t(t-1)} \tag{8}$$

(ii) $$p_i^{AF}(t+1) - p_i^{AF}(t) = \frac{\mathbf{1}[S_t = i][\Xi_{t-1} + 2(t-1)] - (\xi_i + 2)[\xi_i + D_{t-1}(i)]}{(\Xi_{t-1} + 2(t-1))(\Xi_t + 2t)} \tag{9}$$

and $$\mathbb{E}\left[p_i^{AF}(t+1) - p_i^{AF}(t) \mid \mathscr{F}_{t-1}, \xi_t\right] = -\frac{(\xi_i + D_{t-1}(i))(\xi_t + 1)}{(\Xi_{t-1} + 2(t-1))(\Xi_t + 2t)} \tag{10}$$

(iii) $$p_i^{MF}(t+1) - p_i^{MF}(t) = \xi_i \frac{\mathbf{1}[S_t = i]\chi_{t-1} - D_{t-1}(i)(\xi_{S_t} + \xi_t)}{\chi_{t-1}\chi_t} \tag{11}$$

and $$\mathbb{E}\left[p_i^{MF}(t+1) - p_i^{MF}(t) \mid \mathscr{F}_{t-1}, \xi_t\right] \gtrsim \xi_i D_{t-1}(i)\frac{\sum_{j \in V_t} \xi_j D_t(j)[\xi_i - \xi_t - \xi_j]}{\chi_{t-1}^3}. \tag{12}$$

Proof Fix $t = 0, 1, \ldots$, and i in V_t.

Preferential Attachment Model The difference in the visibility of node i in the BA model between time $t + 1$ and t is given as

$$p_i^{BA}(t + 1) - p_i^{BA}(t) = \frac{D_t(i)}{2t} - \frac{D_{t-1}(i)}{2(t-1)}$$

$$= \frac{D_{t-1}(i) + \mathbf{1}[S_t = i]}{2t} - \frac{D_{t-1}(i)}{2(t-1)} \quad (13)$$

and (7) follows. Furthermore, by noting that when looking at the expected difference in visibility conditioned on $\mathscr{F}_{t-1}$, S_t is the only random variable in (13), we obtain

$$\mathbb{E}\left[p_i^{BA}(t + 1) - p_i^{BA}(t) \mid \mathscr{F}_{t-1}\right] = \frac{D_{t-1}(i) + \mathbf{P}[S_t = i \mid \mathbb{G}_{t-1}]}{2t} - \frac{D_{t-1}(i)}{2(t-1)} \quad (14)$$

and (8) follows.

Additive Fitness Model Similarly in the additive fitness model, the difference in the visibility of node i can be written as

$$p_i^{AF}(t + 1) - p_i^{AF}(t)$$

$$= \frac{\xi_i + D_t(i)}{\sum_{j \in V_t} \xi_j + D_t(j)} - \frac{\xi_i + D_{t-1}(i)}{\sum_{j \in V_{t-1}} \xi_j + D_{t-1}(j)}$$

$$= \frac{\xi_i + D_{t-1}(i) + \mathbf{1}[S_t = i]}{\varXi_{t-1} + \xi_t + 2t} - \frac{\xi_i + D_{t-1}(i)}{\varXi_{t-1} + 2(t-1)} \quad (15)$$

and the result (9) follows after simple algebraic manipulations. Taking expectation on both sides conditioned on $\mathscr{F}_{t-1}$ and ξ_t leads to (10).

Multiplicative Fitness Model The difference in the visibility of node i can be written for the multiplicative model as follows

$$p_i^{MF}(t + 1) - p_i^{MF}(t)$$

$$= \frac{\xi_i D_t(i)}{\sum_{j \in V_t} \xi_j D_t(j)} - \frac{\xi_i D_{t-1}(i)}{\sum_{j \in V_{t-1}} \xi_j D_{t-1}(j)}$$

$$= \frac{\xi_i [D_{t-1}(i) + \mathbf{1}[S_t = i]]}{\chi_{t-1} + \xi_{S_t} + \xi_t} - \frac{\xi_i D_{t-1}(i)}{\chi_{t-1}} \quad (16)$$

and the result (11) follows directly. Furthermore, we lower bound the expected change in visibility as follows

$$\mathbb{E}\left[p_i^{MF}(t+1) - p_i^{MF}(t) \mid \mathscr{F}_{t-1}, \xi_t\right]$$

$$= \xi_i \left[\frac{\chi_{t-1}\mathbf{P}\left[S_t = i \mid \mathscr{F}_{t-1}, \xi_t\right]}{\chi_{t-1}\left(\chi_{t-1} + \xi_i + \xi_t\right)} - \frac{D_{t-1}(i)}{\chi_{t-1}(j)} \left[\sum_{\ell \in V_{t-1}} \mathbf{P}\left[S_t = \ell \mid \mathscr{F}_{t-1}, \xi_t\right] \cdot \frac{\xi_\ell + \xi_t}{\chi_{t-1} + \xi_\ell + \xi_t} \right] \right]$$

$$\geq \xi_i \left[\frac{\xi_i D_{t-1}(i)}{\chi_{t-1}\left(\chi_{t-1} + \xi_i + \xi_t\right)} - \frac{D_{t-1}(i)}{\chi_{t-1}} \cdot \frac{\sum_{\ell \in V_{t-1}} \xi_\ell D_{t-1}(\ell)(\xi_\ell + \xi_t)}{\chi_{t-1}^2} \right]$$

$$= \frac{\xi_i D_{t-1}(i)}{\chi_{t-1}} \left[\frac{\xi_i \chi_{t-1}^2 - \sum_{\ell \in V_{t-1}} \xi_\ell D_{t-1}(\ell)(\xi_\ell + \xi_t)[\chi_{t-1} + \xi_i + \xi_t]}{\chi_{t-1}^2\left(\chi_{t-1} + \xi_i + \xi_t\right)} \right]$$

$$\simeq \xi_i D_{t-1}(i) \left[\frac{\xi_i \chi_{t-1} - \sum_{\ell \in V_{t-1}} \xi_\ell D_{t-1}(\ell)(\xi_\ell + \xi_t)}{\chi_{t-1}^2\left(\chi_{t-1} + \xi_i + \xi_t\right)} \right]$$

and the result follows.

From Lemma 2, it is evident that in the BA model, the visibility of the node increases if it forms a new connection. However, in expectation, the visibility does decrease, with the decrease being directly proportional to the degree $D_{t-1}(i)$. The intuition behind this observation is that nodes with high degree already have a high visibility in the network, and therefore their reduction in visibility would be greater than other nodes whose visibility is low to begin with. For the additive fitness model, the visibility increases when a new connection is formed if $\varXi_{t-1} + 2(t-1) > (\xi_t + 2)[\xi_i + D_{t-1}(i)]$. It is expected that the above condition would be satisfied for sufficiently large values of t, unless the fitness value ξ_t, or ξ_i, or both, are very large. For reasons similar to that in the BA model, the visibility decreases in the expected sense in the additive fitness model as well, with the decrease being directly proportional to the sum of degree and fitness values. Also, observe that the decrease in visibility is directly proportional to the fitness of the new node which enters the graph at time t.

On the other hand, from (12) it is evident that, in the multiplicative fitness model, *the visibility could increase in an expected sense* provided the fitness of the node is sufficiently high. This result clearly sets apart the multiplicative fitness model from the other two network models. Furthermore, the expected change in visibility is found to be directly proportional to the product of fitness and the present degree of the node, which allows a greater increase in the visibility of an influential node in comparison to the other network models.

4 Simulation Results

In this section, we present simulation results to compare the visibility of nodes in the BA model with additive and multiplicative fitness models. In all our simulations, we sample fitness values of each incoming node from a Pareto distribution with the following probability distribution $p(t) = \alpha/t^{\alpha+1}$ for $t \geq 1$, and parameter α.

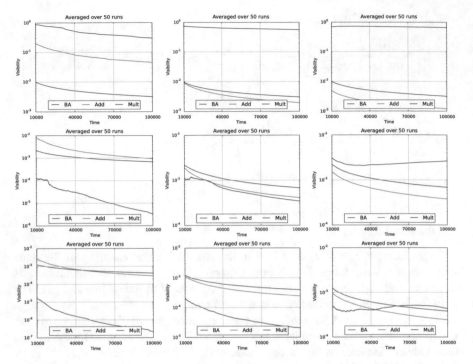

Fig. 1 Visibility of nodes (averaged over 50 independent runs) over time in the three growth models. The columns represent different parameters of the Pareto distribution: $\alpha = 1, 2, 3$. Each row shows visibility results of the kth most influential node after N iterations, where $k = 1, 10, 30$ (*from top to bottom*)

We are interested in studying how the visibility of leaders or influential nodes changes over time in the BA, additive and multiplicative fitness models. To do this, for each given growth model and parameter value α of the Pareto distribution, we first build a graph till $N = 10,000$ nodes using the growth model. We then identify the top k nodes with the highest visibility, and then track the change in the visibility of those k nodes for the next $T = 90,000$ iterations. We average the results over 50 independent runs.

Figure 1 shows average visibility results of the kth most influential node after N iterations, where $k = 1, 10, 30$, and for $\alpha = 1, 2, 3$. From the plots, we observe that, for the topmost node ($k = 1$), the multiplicative fitness model can reach and maintain a higher visibility than the other models for the chosen period of time; which is in agreement with the theoretical analysis given in Lemma 2. Further, notice that while the additive fitness model can reach higher visibility on average than the BA model when $\alpha = 1$, it actually decreases and goes lower than the BA model for $\alpha = 2, 3$. This makes intuitive sense since for $\alpha = 1$, the fitness distribution has a mean that is not finite. Thus, high values of fitness result in some of the nodes being able to attain higher visibility than BA with $\alpha = 1$, while this effect is mitigated for greater values of α. Note that frequent high fitness value of incoming nodes, while helping a node attain high visibility for some time, also results in a faster rate of decrease in visibility once a higher fitness node enters the graph. This is exhibited in the plots where additive fitness models show a faster rate of decrease in visibility than the other models. As predicted in the analysis for the additive model [see (10)], the present scenario demonstrates the decrease in expected visibility due to incoming nodes with high fitness values. However, this effect is less pronounced in the multiplicative model since the degree of the incoming node is small, which smooths out its effect on the visibility of the node being tracked.

Another interesting observation from Fig. 1 is that when α is high, the multiplicative fitness model can maintain a greater number of nodes with high visibility. On the other hand, when $\alpha = 1$, a few very high fitness nodes completely take over and drive down the visibility of other nodes. For $\alpha = 2, 3$ and $k = 10, 30$, the average visibility behavior is seen to increase after an initial decrease, something that is not observed in any plot for the other network models. This suggests that the observed nodes acquire new connections in the mentioned time period, which lead to this dramatic upturn in their visibility.

In Fig. 2, we show some individuals runs for each α setting of the topmost visible ($k = 1$) node. From our individual runs, we notice that runs with $\alpha = 1$ have higher variance, and the variance progressively decreases as α increases. Since lower value of α results in more frequent incoming nodes with high fitness, this leads to higher variance among the individual results. This also explains the sudden drops in visibility of the additive fitness model that we observe in Fig. 1, which is in effect due to a very high fitness incoming node.

5   Conclusion

In this paper, we study the visibility profile of nodes in three network growth models. Firstly, we observe that in the multiplicative fitness model, nodes with high fitness values can successfully maintain visibility in the network to a greater extent when compared with the additive fitness and BA models. High fitness nodes in the additive fitness model, on the other hand, might not always be able to maintain visibility over time even when compared with the BA model. Future work will stress on finding further results on visibility of nodes in the limit of large graphs. Furthermore, it would be interesting to analyse the fraction of nodes that remain influential and reach infinite degree in the limit of large graph size for the fitness models. Also, it is generally accepted and scientifically shown [5, 6, 11] that teams' success dominate individual success, and thus an interesting extension would be to study teams visibility profile rather than individual visibility profile.

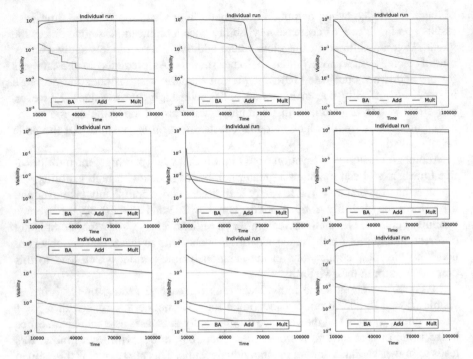

Fig. 2 Change in visibility of the *topmost* ($k = 1$) node in three representative individual runs of the model. The rows correspond to different parameters of the Pareto distribution: $\alpha = 1, 2, 3$ (*from top to bottom*). For each value of α the columns correspond to the visibility patterns in three individual runs

References

1. Barabási, A.L.: Scale-free networks: a decade and beyond. Science **325**(5939), 412–413 (2009). doi:10.1126/science.1173299
2. Barabási, A.L., Albert, R.: Emergence of scaling in random networks. Science **286**(5439), 509–512 (1999)
3. Bianconi, G., Barabási, A.L.: Bose-Einstein condensation in complex networks. Phys. Rev. Lett. **86**(24), 5632 (2001)
4. Ergün, G., Rodgers, G.: Growing random networks with fitness. Phys. A Stat. Mech. Appl. **303**(1), 261–272 (2002)
5. Guimera, R., Uzzi, B., Spiro, J., Amaral, L.A.N.: Team assembly mechanisms determine collaboration network structure and team performance. Science **308**(5722), 697–702 (2005)
6. Jones, B.F., Wuchty, S., Uzzi, B.: Multi-university research teams: shifting impact, geography, and stratification in science. Science **322**(5905), 1259–1262 (2008)
7. Mitzenmacher, M.: A brief history of generative models for power law and lognormal distributions. Internet Math. **1**(2), 226–251 (2004)
8. Newman, M.: Networks: An Introduction. Oxford University Press, New York (2010)

9. Newman, M.E.: The structure and function of complex networks. SIAM Rev. **45**(2), 167–256 (2003)
10. Noulas, A., Shaw, B., Lambiotte, R., Mascolo, C.: Topological properties and temporal dynamics of place networks in urban environments. In: WWW, pp. 431–441 (2015)
11. Wuchty, S., Jones, B.F., Uzzi, B.: The increasing dominance of teams in production of knowledge. Science **316**(5827), 1036–1039 (2007)

Topological Data Analysis of Critical Transitions in Financial Networks

Marian Gidea

Abstract We develop a topology data analysis-based method to detect early signs for critical transitions in financial data. From the time-series of multiple stock prices, we build time-dependent correlation networks, which exhibit topological structures. We compute the persistent homology associated to these structures in order to track the changes in topology when approaching a critical transition. As a case study, we investigate a portfolio of stocks during a period prior to the US financial crisis of 2007–2008, and show the presence of early signs of the critical transition.

Keywords Stock correlation network • Critical transition • Topological data analysis

1 Introduction

A *critical transition* refers to an abrupt change in the behavior of a complex system—arising due to small changes in the external conditions—which makes the system switch from one steady state to some other steady state, after undergoing a rapid transient process (e.g., 'blue-sky catastrophe' bifurcation). Examples of critical transitions are ubiquitous, including market crashes, abrupt shifts in ocean circulation and climate, regime changes in ecosystems, asthma attacks and epileptic seizures, etc. A landmark paper on the theory of critical transitions and its applications is [19].

A challenging problem of practical interest is to detect *early signs* of critical transitions, that is, to identify significant changes in the structure of the time-series data emitted by the system prior to a sharp transition. In this paper we propose a new method to look for critical transitions, based on measuring changes in the topological structure of the data. We consider systems that can be described as time-varying weighted networks, and we track the changes in the topology of the network as the system approaches a critical transition. We use tools from *topological data*

M. Gidea (✉)
Department of Mathematical Sciences, Yeshiva University, New York, NY 10016, USA
e-mail: Marian.Gidea@yu.edu

© Springer International Publishing AG 2017
E. Shmueli et al. (eds.), *3rd International Winter School and Conference on Network Science*, Springer Proceedings in Complexity, DOI 10.1007/978-3-319-55471-6_5

analysis, more precisely *persistent homology*, to provide a precise characterization of the topology of the network throughout its time-evolution. We observe, in empirical data, that there are significant, measurable changes in the topology of the network as the underlying system approaches a critical transition.

The pipeline of our approach is the following. The *input* of our procedure is a time-evolving weighted network $G(V, E)$, $w_t : E \rightarrow [0, \infty)$, i.e., a graph of nodes V and edges E, with each edge $e \in E$ having assigned a weight $w_t(e)$ which varies in time. At each instant of time t, using a threshold value of the weight function as a parameter, we consider the threshold sub-network consisting of those edges whose weights are below that threshold. We compute the homology of the clique complex determined by that sub-network. As we vary the threshold value, some of the homology generators persist for a large range of values while others disappear quickly. The persistent generators provide information about the significant, intrinsic patterns within the network, while the transient patterns may be redeemed as less significant or random. This information can be encoded in terms of a so-called *persistent diagram*, which provides a summary of the topological information on the network. As the time evolves, the topology of the network changes, and the corresponding persistent diagrams also change. There is a natural metric (in fact, several) to measure distances between persistent diagrams. It is important to note that *persistent diagrams are robust*, meaning that small changes in the network yield persistent diagrams that are close to one another in terms of their mutual distances. The *output* of our procedure consists of a sequence of distances measured between the persistent diagram at time t and the persistent diagram at some initial time t_0.

The salient features of our approach are the following:

(i) We process the input signal in its entirety, as we do not filter out noise from signal,
(ii) For weighted networks, we obtain a global description of all threshold sub-networks, for all possible threshold values;
(iii) We describe in more detail the structures of our networks, unlike the statistical-type methods (e.g., centrality measures);
(iv) We provide an efficient way to compare weighted networks through the distances between the associated persistent diagrams,
(v) For time-dependent networks, we track the changes of the topology of the networks via the distances between persistent diagrams.

We point out that the networks that we consider in this paper are very noisy. Metaphorically speaking, what we are trying to do here is to quantify the 'shape of noise'.

We illustrate our procedure by investigating financial time series for the US financial crisis of 2007–2008. The time-varying network that we consider is the cross correlation network $C = (c_{i,j})$ of the stock returns for the companies in the Dow Jones Industrial Index (DJIA); the nodes of the network represent the stocks, and the weights of the edges are given by the distances $d_{i,j} = \sqrt{2(1 - c_{i,j})}$. Following the process described above, we compute the time series of the distances between the persistent diagrams at time t and the reference persistent diagram at initial time t_0. The conclusion is that these time series display a significant change prior to the critical transition (i.e., the peak of the crisis), which shows that the stock correlation network undergoes significant changes in its topological structure.

For the computation of persistent diagrams and their mutual distances we use the R package *TDA* [10].

2 Background

We provide a brief, largely self-contained, review of the persistent homology method, and describe how to use it to analyze the topology of weighted networks. Some general references and applications include [1–3, 6, 8, 13, 14, 16].

2.1 Persistent Homology

Persistent homology is a computational method to extract topological features from a given data set (e.g., a point-cloud data set or a weighted network) and rank them according to some threshold parameter (e.g., the distance between data points or the weight of the edges). Topological features that are only visible at low levels of the parameter are ranked lower than topological features that are visible at both low and high levels. For each value of the threshold parameter one builds a simplicial complex (i.e., a space made from simple pieces—geometric simplices, which are identified combinatorially along faces). In our case, the vertices correspond to the data points and the simplices are determined by the proximity of data points. When the threshold parameter is varied, the corresponding simplicial complexes form a filtration (i.e., an ordering of the simplicial complexes that is compatible with the ordering of the threshold values). Then one tracks the topological features (e.g., connected components, 'holes' of various dimensions) of the simplicial complexes across the filtration, and record for each topological feature the value of the parameter at which that feature appears for the first time ('birth value'), and the value of the parameter at which the feature disappears ('death value'). We now give technical details.

A simplicial complex K is a set of simplices $\{\sigma\}$ of various dimensions that satisfies the following conditions: (1) any face of a simplex $\sigma \in K$ is also in K, and (2) the intersection of any two simplices $\sigma_1, \sigma_2 \in K$ is either $\emptyset$ or a face of both σ_1 and σ_2.

Given a simplicial complex K, we denote by $H_i(K)$ the ith homology group with coefficients in $\mathbb{Z}_2$. This is a free abelian group whose generators consists of certain chains of i-dimensional simplices (i.e., cycles that are not boundaries). Note that $H_i(K) = 0$ for $i \geq m + 1$. The generators of the ith homology group account for the 'independent holes' in K at dimension i. For example, the number of 0-dimensional generators equals that of connected components of K, the number of 1-dimensional generators equals that of 'tunnels' (or 'loops'), the number of 2-dimensional generators equals that of 'cavities', etc. For a reference, see, e.g., [12].

A *filtration* of K is a mapping $a \in A \mapsto \mathscr{F}(a) := K_a \subseteq K$, from a (totally ordered) set of parameter values $A \subseteq \mathbb{R}$ to a set of simplicial sub-complexes of K, satisfying the *filtration condition*: $a \leq a' \Rightarrow K_a \subseteq K_{a'}$. For any filtration of simplicial complexes $a \mapsto K_a$ the corresponding homology groups also form a filtration $a \mapsto H_i(K_a)$, that is, $a \leq a' \Rightarrow H_i(K_a) \subseteq H_i(K_{a'})$.

For $a \leq a'$, the inclusion $H_i(K_a) \subseteq H_i(K_{a'})$ induces a group homomorphism $h_i^{a,a'} : H_i(K_a) \to H_i(K_{a'})$, in all i. Let $H_i^{a,a'} = \mathrm{Im}(h_i^{a,a'})$ be the image of $h_i^{a,a'}$ in $H_i(K_{a'})$. We say that a homology class $\gamma \in H_i(K_b)$ is *born* at the parameter value $a = b$ if $\gamma \notin H_i^{b-\delta,b}$ for any $\delta > 0$. If γ is born at K_b then we say that it *dies* at the parameter value $a = d$, with $b \leq d$, if γ coalesces with an older class in $H_i(K_{d-\varepsilon})$ as we go from $K_{d-\varepsilon}$ to K_d for $\varepsilon > 0$, that is, $h_i^{b,d-\varepsilon}(\gamma) \notin H_i^{b-\delta,d-\varepsilon}$ for any small $\varepsilon, \delta > 0$, but $h_i^{b,d}(\gamma) \in H_i^{b-\delta,d}$ for some small $\delta > 0$. If γ is born at K_d but never dies then we say that it dies at infinity. Thus, we have a value $b(\gamma) = b$ and a death value $d(\gamma) = d$ for each generator γ that appears in the filtration of homology groups. The persistence, or 'life span' of the class γ is the difference between the two values, $\mathrm{pers}(\gamma) = d(\gamma) - b(\gamma)$.

The ith persistent diagram of the filtration $\mathscr{F}$ is defined as a multiset P_i in $\mathbb{R}^2$, for $i = 0, \ldots, m$, obtained as follows:

- For each class γ_i we assign a point $z_i = (b_i, d_i) \in \mathbb{R}^2$ together with a multiplicity $\mu_i(b_i, d_i)$; where b_i is the parameter value when γ_i is born, and d_i is the parameter value when γ_i dies. The multiplicity $\mu_i(b_i, d_i)$ of the point $z_i = (b_i, d_i)$ equals the number of classes γ_i that are born at b_i and die at d_i. This multiplicity is finite since the simplicial complex is finite.
- In addition, P_i contains all points in the positive diagonal of $\mathbb{R}^2$. These points represent all trivial homology generators that are born and die at every level. Each point on the diagonal has infinite multiplicity.
- The axes of the persistent diagram are birth values on the horizontal axis and death values on the vertical axis.

An illustration of persistent diagrams for a simple example of point-cloud data set and a filtration of simplicial complexes is shown in Fig. 1.

The space of persistent diagrams can be endowed with a metric space structure. A standard metric that can be used is the *degree p Wasserstein distance* (earth mover distance), with $p > 0$. This is defined by $D_p(P_i^1, P_i^2) = \inf_{\phi} \left[\sum_{q \in P_i^1} \|q - \phi(q)\|_{\infty}^p \right]^{1/p}$,

where the summation is over all bijections $\phi : P_i^1 \to P_i^2$. When $p = \infty$ the Wasserstein distance D_{∞} is known as the bottleneck distance. Since the diagonal set is by default part of all persistent diagrams, the pairing of points between P_i^1 and P_i^2 via ϕ can include pairings between off-diagonal points and diagonal points.

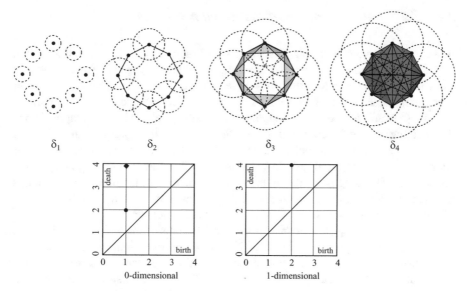

Fig. 1 A point-cloud data set representing a 'noisy' circle, together with a filtration of simplicial complexes, corresponding to some threshold parameter values $\delta_1 < \delta_2 < \delta_3 < \delta_4$. The 0-dimensional and the 1-dimensional persistence diagrams are shown at the *bottom* of the figure. At δ_1 there are eight connected components and no 1-dimensional hole. At δ_2 the eight connected components coalesce into a single one, indicated by the point $(1, 2)$ in the 0-dimensional diagram, which has multiplicity 7; also, a 1-dimensional hole is born. There is no topological change at δ_3. At δ_4 the 1-dimensional hole gets filled in and dies, indicated by the the point $(2, 4)$ in the 1-dimensional diagram; the single connected component continues living for ever, and is represented by *filled diamond*

We note that different value of the degree p yield different types of measurements of the distances between persistent diagrams. Using $p = \infty$, the corresponding distance D_∞ only measures the distance between the most significant features (farthest from the diagonal) in the diagrams, matched via some appropriate ϕ. Using $p \geq 1$ large, the corresponding distance D_p puts more weight on the significant features (farther from the diagonal) than on the least significant ones (closer to the diagonal). Using $p > 0$ small has just the opposite effect on the measurement.

One of the remarkable properties of persistent diagrams is their robustness, meaning that small changes in the initial data produce persistent diagrams that are close to one another relative to Wasserstein metric. The essence of the stability results is that the persistent diagrams depend Lipschitz-continuously on data. For details see [5, 7, 9].

2.2   Persistent Homology of Weighted Networks

A weighted network is a pair consisting of a graph $G = G(V, E)$ and a weight function associated to its edges $w : E \to [0, +\infty)$; let $\theta_{max} = \max(w)$. In the sequel we will only consider graphs that are simple and undirected. In examples, the weight function is chosen so that nodes with similar characteristics are linked together.

One standard recipe to investigate the topology of weighted graphs is via thresholding, that is, by considering only those edges whose weights are below (or above) some suitable threshold, and study the features of the resulting graph. Of course, the choice of the threshold value makes a difference in the topology of the resulting graph. Using persistent homology, we can extract the topological features for each threshold graph, and represent all these features, ranked according to their 'life span', in a persistent diagram. We now give technical details.

For each $\theta \in [0, \theta_{max}]$, we consider the sub-level sets of the weight function, that is, we restrict to subgraphs $G(\theta)$ which keep all edges of weights w below or equal to the threshold θ. The graphs obtained by restricting to successive thresholds have the filtration property, i.e., $\theta \le \theta' \implies G(\theta) \subseteq G(\theta')$. In a similar way, we can consider super-level sets, by restricting to subgraphs $G(\theta)$ which keeps all edges of weights w above or equal to the threshold θ. Super-level sets can be thought of as sub-level sets of the weight function $w' = \theta_{max} - w$.

For each threshold graph $G(\theta)$ we construct the *Rips complex* (clique complex) $K = X(G(\theta))$. This is defined as the simplicial complex with all complete subgraphs (cliques) of $G(\theta)$ as its faces. That is, the 0-skeleton of K consists of just the vertices of $G(\theta)$, the 1-skeleton of all vertices and edges—which is the graph $G(\theta)$ itself—the 2-skeleton of all vertices, edges, and filled triangles, etc. High dimensional cliques correspond to highly interconnected clusters of nodes with similar characteristics (as encoded by the weight function). The filtration of the threshold subgraphs yields a corresponding filtration of the Rips complexes $\theta \mapsto K_\theta := X(G(\theta))$; thus, $\theta \le \theta' \implies K_\theta \subseteq K_{\theta'}$. As it was noted before, the homology groups associated to this filtration satisfy themselves the filtration property, i.e., $\theta \le \theta' \implies H_i(K_\theta) \subseteq H_i(K_{\theta'})$. From this point on, we can compute the persistent homology and the persistent diagrams associated to this filtration, in the manner described in Sect. 2.1.

In Sect. 3 we will only compute persistent diagrams of dimension 0 and 1, so we explain in detail the significance of these diagrams in terms of the threshold networks.

A point (θ_b, θ_d) in a 0-dimensional persistent diagram has the following meaning:

- At the threshold value θ_b a connected component is born, where each pair of nodes in the component is connected via a path of edges of weights $\theta \le \theta_b$;
- At the threshold value θ_d two or more connected components coalesce into a single one, via the addition of one or several edges of weight $\theta = \theta_d$ to the threshold network.

A point (θ_b, θ_d) in a 1-dimensional persistent diagram has the following meaning:

- At the threshold value θ_b a loop of 4 or more nodes is born, whose nodes are connected in circular order by edges of weights $\theta \leq \theta_b$; note that a loop of 3 nodes yields a complete sub-graph in the Rips complex (i.e., a filled triangle), which carries no 1-dimensional homology;
- At the threshold value $\theta = \theta_d$ one or more loops get covered by filled triangles, due to adding one or more edges of weight $\theta = \theta_d$, thus making the corresponding 1-dimensional homology generators vanish.

We note that applications of persistent homology to networks also appear, e.g., in [4, 11].

3 Detection of Critical Transitions from Correlation Networks

3.1 Correlation Networks as Weighted Networks

The network that we analyze here is derived from the DJIA stocks listed as of February 19, 2008. We utilize the time series of the daily returns based on the adjusted closing prices $S_i(t)$, i.e., $x_i(t) = \frac{S_i(t+\Delta t)-S_i(t)}{S_i(t)}$, where $\Delta t = 1$ day, and the indices i correspond to the individual stocks. We restrict to the data from January 2004 to September 2008 (when Lehman Brothers filed for bankruptcy).

Now we define the weighted network $G(V, E)$ that we analyze. The vertices V of the network correspond to the individual DJIA stocks. Each pair of distinct vertices $i, j \in V$ is connected by an edge e, and each edge is assigned a weight $w(e, t)$ at time t defined as follows:

- Compute the Pearson correlation coefficient $c_{i,j}(t)$ between the nodes i and j at time t, over a time horizon T, by $c_{i,j}(t) = \frac{\sum_{\tau=t-T}^{t}(x_i(\tau)-\bar{x}_i)(x_j(\tau)-\bar{x}_j)}{\sqrt{\sum_{\tau=t-T}^{t}(x_i(\tau)-\bar{x}_i)^2}\sqrt{\sum_{\tau'=t-T}^{t}(x_j(\tau')-\bar{x}_j)^2}}$, where $\bar{x}_i, \bar{x}_j$ denote the averages of $x_i(t), x_j(t)$ respectively, over the time interval $[t-T, t]$;
- Compute the distance between the nodes i and j, $d_{i,j}(t) = \sqrt{2(1-c_{i,j}(t))}$—the fact that the metric axioms are satisfied follows easily from the properties of correlation;
- Assign the weight $w(e, t) = d_{i,j}(t)$ to the edge e between i and j.

For the computation of the correlation via the Pearson estimator, there is empirical evidence against using longer time horizons when non-stationary behavior is present. Therefore, in our computation we use a rather short time horizon of $T = 15$; we also use the arithmetic return rather than the standard log return. For an argument in support of these choices see [15].

The range of values of $d_{i,j}$ is $[0, 2]$. Note that $d(i, j) = 0$ if the nodes i and j are perfectly correlated, and $d(i, j) = 2$ if the nodes are perfectly anti-correlated. Edges between correlated nodes have smaller weights, and edges between uncorrelated/anti-correlated nodes have bigger weights. Since correlation between stocks is of interest, we focus on edges with low values of d.

In the sequel, we will consider both sub-levels sets and super-level sets of the weight function.

Each sub-level set of the weight function w, at a threshold level $\theta \in [0, 2]$, yields a subgraph $G(\theta)$ containing only those edges for which $0 \le d_{i,j} \le \theta$, that is, $G(\theta) = \{e = e(i, j) \mid 1 - \frac{1}{2}\theta^2 \le c_{i,j} \le 1.\}$ When θ is small, $G(\theta)$ contains only edges between highly-correlated nodes. As θ is increased up to the critical value $\sqrt{2} = 1.414214$ edges between low-correlated nodes are progressively added to the network. As θ is increased further, edges between anti-correlated nodes appear in the network.

Each super-level set of the weight function $w = d$ can be conceived as a sub-level sets for the dual weight function $w' = 2 - d$. The sub-level set $G(\theta)$ for this weight-function contains only those edges for which $d_{i,j} \ge 2 - \theta$, hence $G_{w'}(\theta) = \{e = e(i, j) \mid -1 \le c_{i,j} \le 1 - \frac{1}{2}(2 - \theta)^2.\}$ When θ is small, $G_{w'}(\theta)$ contains only edges between anti-correlated nodes. When θ crosses the critical value $2 - \sqrt{2} = 0.5857864$, edges between low-correlated nodes are progressively added to the network. As θ is increased further towards the highest possible value of 2, highly-correlated nodes are added to the network.

Sub-level sets and super-level sets produce very different type of networks, and they furnish complementary information. We will discuss this in Sect. 3.2.

3.2   Persistent Diagrams of Correlation Networks

In this section we use persistent homology to quantify the changes in the topology of the correlation networks when approaching a critical transition. For illustrative purposes, we show some correlation networks in Fig. 2; the top three networks represent instants of time far from the beginning of the 2007–2008 financial crisis, while the bottom three diagrams represent instants of time preceding the crisis.

We compute persistent homology in dimensions 0 and 1 for the correlation network from Sect. 3.1. We do not consider higher-dimensional persistent homology because the network is very small, so the presence higher-dimensional cliques is likely accidental.

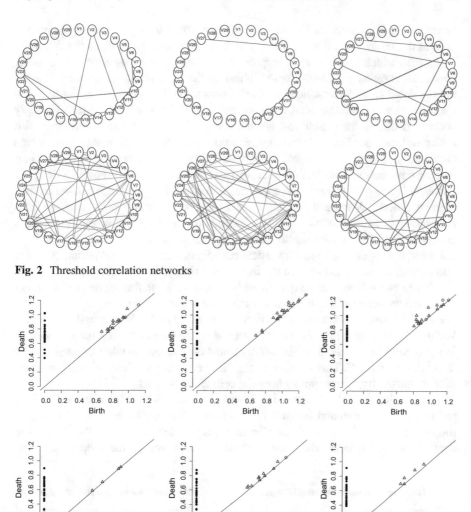

Fig. 2 Threshold correlation networks

Fig. 3 Persistent diagrams (sub-level sets)

First, we consider threshold networks given by the sublevel sets of the weight function $w = d$. Several persistent diagrams are shown in Fig. 3. The top three diagrams correspond to instants of time far from the beginning of the 2007–2008 financial crisis, while the bottom three diagrams correspond to instants of time preceding the crisis.

The 0-dimensional persistent homology provides information on how the network connectivity changes as the value of θ in increased from 0 to 2. Each black dot on the persistent diagram corresponds to one (or several) connected component of the graph. The horizontal coordinate of each dot is 0, since all components are born at threshold value $\theta = 0$. The vertical coordinate of a dot gives the threshold value θ at which a connected component dies, by joining together with another connected component. The dot with highest vertical coordinate (other than 2) gives the threshold value θ for which the graph becomes fully connected. A dot at 2 (the maximum value) indicates that once the graph is fully connected, it remains fully connected (hence the component never dies) as θ is further increased. Dots with lower vertical coordinates indicate threshold values for which smaller connected components consisting of highly correlated nodes die, i.e., coalesce together into larger components. Dots with higher vertical coordinates correspond to death of connected components due to the appearance of edges between uncorrelated or anti-correlated nodes. We recall that the critical value of θ that marks the passage from correlation to anti-correlation is 1.41. Inspecting the diagrams in Fig. 3 we see a concentration of dots with higher vertical coordinates in the first period, and a a concentration of dots with lower vertical coordinates in the second period. There is less correlation in the network in the first period than in the second period.

These observations can be quantified by computing the time-series representing the distances between the diagram at time t and some reference persistent diagram at the initial time t_0. We sample this time series at $\Delta t = 10$. We show this in Fig. 4. We use the Wasserstein distance of degree $p = 2$. We notice that the oscillations in the second half of the time series are almost double in size when compared with those in the first half. This shows a change in the topology of the network, in terms of its connectivity, when approaching the critical transition.

Now we interpret the 1-dimensional persistent homology, illustrated in Fig. 3 by red marks. The horizontal coordinate of a mark gives the birth value of a loop in the network, and the vertical coordinate gives the death value of that loop. The death of a loop occurs when edges between the nodes of the loop appear and

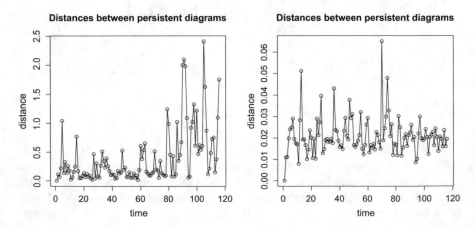

Fig. 4 *Left*: distances between 0-dimensional persistent diagrams (sub-level sets). *Right*: distances between 1-dimensional persistent diagrams (sub-level sets)

form complete 2-simplices (filled triangles) that fill up the loop. Dots with low coordinates indicate the presence of cliques that are highly correlated. Marks with higher vertical coordinates indicate the death of loops due to edges between low-correlated or anti-correlated nodes. The top three diagrams in Fig. 3 seem to indicate a concentration of dots at a higher range of values when compared with the bottom three diagrams.

We also compute the time-series (sampled at $\Delta t = 10$) of the Wasserstein distances of degree $p = 2$, between the diagram at time t and the reference diagram at t_0. We show this in Fig. 4. The oscillations in the second part of the time series are smaller than the ones in the first part. Again, there is a change in the topology of the network, in terms of its cliques, when approaching the critical transition: the number of loops of correlated nodes appears to stabilize itself.

We now compute the super-levels sets of w, which are sub-level set of w'. The resulting persistent diagrams have a different interpretation. The critical value of the threshold θ for the switch from anti-correlation to correlations is 0.5857864. Points in the persistent diagram with low vertical coordinates correspond to anti-correlation/non-correlation, and points with higher value of the vertical coordinate (other than 2) indicates the appearance of edges between correlated nodes. A point on the persistent diagram with higher vertical coordinate represents the death of a connected component (or a loop), possibly formed by anti-correlated or low correlated nodes, when an edge between correlated nodes is added to the networks. Thus, the homology generators identified by the persistent diagrams represent cliques of stocks associated to 'normal' market conditions (which are associated to lack of correlation). The death of these generators is caused by the addition to correlated edges to the threshold network (in dimension 0, by joining together different connected components, and in dimension 1 by closing the loops). That is, the persistence diagrams capture the loss of normal market conditions. We show some persistent diagrams in Fig. 5, the time series of distances between 0-dimensional persistent diagrams, and between 1-dimensional persistent diagrams, in Fig. 6.

4 Conclusions

The analysis of the persistent diagrams and of the distances between persistent diagrams show significant changes in the topology of the correlation network in the period prior to the onset of the 2007–2008 financial crisis; early signs become apparent starting February 2007 (note that the U.S. stock market peaked in October 2007). These topological changes can be characterized by an increase in the cross correlations between various stocks, as well as by the emergence of sub-networks of cross correlated stocks.

These results are in agreement with other research asserting that crises are typically associated with rapid changes in the correlation structure and in the network topology (see, e.g., [15, 17, 18, 20]).

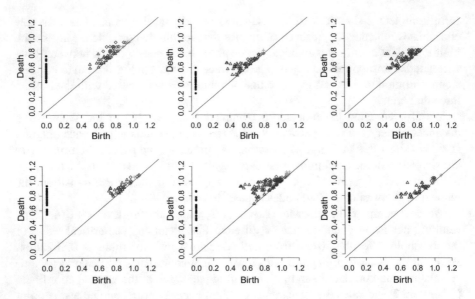

Fig. 5 Persistent diagrams (super-level sets)

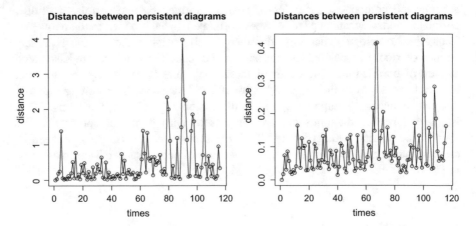

Fig. 6 *Left*: distances between 0-dimensional persistent diagrams (super-level sets). *Right*: distances between 1-dimensional persistent diagrams (super-level sets)

In addition to the experiments presented here, we have used persistent homology to analyze the time-series of some market indices (e.g., the VIX index) for the same period, using point-cloud data sets obtained via delay-coordinate reconstruction method. These tests also show early signs of critical transition; those results will be presented elsewhere.

Acknowledgements Research of Marian Gidea was partially supported by the Alfred P. Sloan Foundation grant G-2016-7320, and by the NSF grant DMS-0635607.

References

1. Adler, R.J., Bobrowski, O., Borman, M.S., et al.: Persistent homology for random fields and complexes. Inst. Math. Stat. Collect. **6**, 124–143 (2010)
2. Berwald, J., Gidea, M.: Critical transitions in a model of a genetic regulatory system. Math. Biol. Eng. **11**(4), 723–740 (2014)
3. Berwald, J., Gidea, M., Vejdemo-Johansson, M.: Automatic recognition and tagging of topologically different regimes in dynamical systems. Discontinuity Nonlinearity Complex. **3**(4), 413–426 (2015)
4. Carstens, C.J., Horadam, K.J.: Persistent homology of collaboration networks. Math. Probl. Eng. **2013**, 1–7 (2013)
5. Chazal, F., Cohen-Steiner, D., Guibas, L.J., et al.: Gromov-Hausdorff stable signatures for shapes using persistence. In: Computer Graphics Forum (Proc. SGP 2009), pp. 1393–1403 (2009)
6. Chazal, F., de Silva, V., Oudot, S.: Persistence stability for geometric complexes. Geom. Dedicata **173**, 193–214 (2013). doi:10.1007/s10711-013-9937-z
7. Cohen-Steiner, D., Edelsbrunner, H., Harer, J., et al.: Lipschitz functions have L_p-stable persistence. Found. Comput. Math. **10**, 127–139 (2010). doi:10.1007/s10208-010-9060-6
8. Edelsbrunner, H., Harer, J.: Persistent homology — a survey. In: Goodman, J.E., Pach, J., Pollack, R. (eds.) Surveys on Discrete and Computational Geometry. Twenty Years Later. Contemporary Mathematics, vol. 453, pp. 257–282. American Mathematical Society, Providence, RI (2008)
9. Edelsbrunner, H., Morozov, M.: Persistent homology: theory and practice. In: European Congress of Mathematics, Krakow, 2–7 July 2012, pp. 31–50. European Mathematical Society, Zürich (2012)
10. Fasy, B.T., Kim, J., Lecci, F., Maria, C.: Introduction to the R package TDA, arXiv:1411.1830 (2014)
11. Horak, D., Maletic, S., Rajkovic, M.: Persistent homology of complex networks. J. Stat. Mech. Theory Exp. **2009**, 1–25 (2009)
12. Kaczynski, T., Mischaikow, K., Mrozek, M.: Computational Homology. Springer, New York (2004)
13. Khasawneh, F., Munch, E.: Chatter detection in turning using persistent homology. Mech. Syst. Signal Process. **70–71**, 527–541 (2016)
14. Kramár, M., Levanger, R., Tithof, J., et al.: Analysis of Kolmogorov flow and Rayleigh-Bénard convection using persistent homology. Phys. D Nonlinear Phenom. **334**, 82–98 (2016). ISSN 0167-2789, http://dx.doi.org/10.1016/j.physd.2016.02.003
15. Münnix, M.C., Shimada, T., Schäfer, R., et al.: Identifying states of a financial market. Sci. Rep. **2**(644), 1–6 (2012)
16. Nicolau, M., Levine, A.J., Carlsson, G.: Topology based data analysis identifies a subgroup of breast cancers with a unique mutational profile and excellent survival. Proc. Natl. Acad. Sci. **108**(17), 7265–7270 (2011)
17. Nobi, A., Sungmin, L., Kim, D.H., et al.: Correlation and network topologies in global and local stock indices. Phys. Lett. A **378**(34), 2482–2489 (2014)
18. Nobi, A., Maeng, S.E., Ha, G.G., et al.: Effects of global financial crisis on network structure in a local stock market. Phys. A Stat. Mech. Appl. **407**, 135–143 (2014)
19. Scheffer, M., Bascompte, J., Brock, W.A., et al.: Early-warning signals for critical transitions. Nature **461**(3), 53–59 (2009)
20. Smerlak, M., Stoll, B., Gupta, A., Magdanz, J.S.: Mapping systemic risk: critical degree and failures distribution in financial networks. PLoS ONE **10**(7), e0130948 (2015). doi:10.1371/journal.pone.0130948

Modeling and Analysis of Glass Ceiling and Power Inequality in Bi-populated Societies

Chen Avin, Zvi Lotker, Yinon Nahum, and David Peleg

Abstract This paper attempts to rigorously analyze the social effects of power inequality and glass ceiling in a society with two populations (e.g. men and women). To this end, we define a mathematical model based on a social network with two populations, in which these phenomena are studied. We define measures for Normalized (or Differential) Power Inequality and Full-Spectrum Glass Ceiling, and formalize the conditions for their existence in terms of three societal parameters, the relative size of the two populations, the level of homophily, and the extent of the "leaky pipeline" phenomenon.

Keywords Glass ceiling • Networks • Model

1 Introduction

Background and Motivation The Federal Glass Ceiling Commission defines Glass Ceiling as "the unseen, yet unbreachable barrier that keeps minorities and women from rising to the upper rungs of the corporate ladder, regardless of their qualifications or achievements" [6]. Glass Ceiling proves to be more prominent in some fields than in others. In particular, it is known that women have been historically underrepresented in the fields of Science, Technology, Engineering and Mathematics (STEM) [2, 3, 5, 7, 8].

C. Avin • Z. Lotker
Ben-Gurion University of the Negev, Be'er Sheva, Israel
e-mail: avin@cse.bgu.ac.il; zvilo@cse.bgu.ac.il

Y. Nahum (✉) • D. Peleg
Weizmann Institute of Science, Rehovot, Israel
e-mail: yinon.nahum@weizmann.ac.il; david.peleg@weizmann.ac.il

© Springer International Publishing AG 2017   61
E. Shmueli et al. (eds.), *3rd International Winter School and Conference on Network Science*, Springer Proceedings in Complexity, DOI 10.1007/978-3-319-55471-6_6

The glass ceiling effect is in fact part of a wider phenomenon, namely, the inequality between men and women[1] in society. In this paper we aim at formalizing the Glass Ceiling effect and analyzing the conditions for its formation. While doing so, we address also several related but distinct *effects* of societal power inequality, as well as several distinct *causes* for such inequality, and attempt to formally analyze the relationships between them.

Inequality Effects The inequality phenomena discussed in this paper relate directly to some measures of power. Given any concrete sub-domain of societal activity (such as politics, art, academia or the economy), we measure the power level of an individual v at a given time t by some function $p(v, t)$. We view this parameter as a random variable dependent on time, on the behavior of other individuals, and on external parameters of the setting. Therefore, our focus is on the expected power, $\mathbb{E}[p(v, t)]$. For a group W of individuals, we may consider the *average expected power*, defined as $\bar{p}(W, t) = \sum_{v \in W} \mathbb{E}[p(v, t)] / |W|$.

The first effect we look at is *Power Inequality*. For two individuals u and w, we may quantify the inequality between them by comparing the values $\mathbb{E}[p(u, t)]$ and $\mathbb{E}[p(w, t)]$. A formal measure for their power inequality could be the difference or the ratio between these two parameters. For two sub-populations R and B (hereafter, the "red" and "blue" communities, intuitively thinking of R as the disadvantaged group), we may similarly compare $\bar{p}(R, t)$ and $\bar{p}(B, t)$. For example, a possible evidence for the existence of power inequality against the red population could be that $\bar{p}(R, t)$ is significantly smaller than $\bar{p}(B, t)$. We say that there is *Power Inequality* for the red community at time t if $\bar{p}(R, t) < \bar{p}(B, t)$. The resulting measure for power inequality[2] is

$$\mathsf{PI}(t) = \bar{p}(R, t)/\bar{p}(B, t). \tag{1}$$

Note that whenever $\lim_{t \to \infty} \bar{p}(B, t) = \infty$ and $\lim_{t \to \infty} \mathsf{PI}(t) < 1$, then not only is there power inequality, but also $|\bar{p}(B, t) - \bar{p}(R, t)| \to \infty$. This leads to our next definition of a strong power inequality. We say that there is a *Strong Power Inequality* for the red community if there exists a constant $c < 1$ such that $\mathsf{PI}(t) \leq c$ for large enough t.

Popular discussions in the media on the Glass Ceiling effect often digress to arguments related to power inequality. Note, however, that while the two effects fall under the same general societal problem, power inequality effects are inherently different from, and independent of, the Glass Ceiling effect. In particular, two sub-populations may (at least theoretically) exhibit Power Inequality with no Glass Ceiling whatsoever (e.g., women may reach the highest echelons in society, and still have lower average power as a group), and vice versa. How should we then define formally the Glass Ceiling effect?

---

[1]Much of the discussion in this paper applies also to minorities; however, for simplicity of presentation we will henceforth restrict the discussion to gender, and denote the two sub-populations as the "red" (women) and "blue" (men) communities, R and B.

[2]Throughout, we interpret $x/0 = \infty$ for $x > 0$.

This question was tackled in [1], where two measures were proposed. The first, termed *Tail Glass Ceiling*, focused on counting the number of highly powerful individuals in each subgroup. The second measure proposed in [1], termed *Moment Glass Ceiling*, relied on a quantity similar to the average power, except that it averaged the *variance* (or, the second moment) of the power parameters.

The two measures of [1] do capture the global behavior of Glass Ceiling in the limit, but still suffer from some limitations. First, those definitions only consider the system behavior in the limit (as t tends to ∞), and do not apply in settings where individuals have a bounded horizon of activity (from hiring to retirement). Second, they only measure the Glass Ceiling as a *global* effect, reflecting on *entire* sub-populations, and do not provide us with a means for directly comparing two individuals. In particular, they do not take into account the personal capabilities and experience of each individual.

In contrast, in this paper we also attempt to identify Glass Ceiling effects that occur on a *local* or individual level. Our approach is motivated by the observation that the power of any individual relies on subjective parameters such as the individual's personal capabilities and life/work experience. These parameters bound the individual's potential. Moreover, different individuals may invest different amounts of time and efforts in developing their careers, and this may significantly influence their power as well. Hence one may rightfully maintain that each individual has a "private" (or "personal") ceiling blocking his or her progress.[3] It follows that in order to justify claims for the existence of Glass Ceiling effects on an individual level, it is necessary to make our comparisons between individuals of similar backgrounds, in order to factor out these subjective parameters. In fact, common operational definitions of Glass Ceiling (cf. [4]) include the characteristics that a Glass Ceiling inequality represents "A gender or racial difference that is not explained by other job-relevant characteristics of the employee."

To handle this issue, we propose a *Normalized* (or *Differential*) Power Inequality measure, denoted NPI, based on comparing groups of individuals of similar capabilities and experience. Abstractly capturing the capabilities and experience of an individual v by a "profile" parameter $\mathcal{P}(v)$, we are interested in comparing $\mathbb{E}\left[p(u,t) \mid \mathcal{P}(u) = P\right]$ with $\mathbb{E}\left[p(w,t) \mid \mathcal{P}(w) = P\right]$ for two individuals u and w. Thus we define a measure

$$\mathsf{NPI}(u, w, t, P) = \frac{\mathbb{E}\left[p(u,t) \mid \mathcal{P}(u) = P\right]}{\mathbb{E}\left[p(w,t) \mid \mathcal{P}(w) = P\right]}, \tag{2}$$

where the expectation is taken over the profiles and behavior of other individuals and possibly other random factors. For a profile P, we say that a *Local Glass Ceiling* effect is formed for R under the profile P if $\mathsf{NPI}(u, w, t, P) < 1$ for all $u \in R, w \in B$. We say that a *Full-Spectrum* (or *Total*) *Glass Ceiling* effect is formed for R if a Local Glass Ceiling effect is formed for *every* profile P, that is, $\mathsf{NPI}(u, w, t, P) < 1$ for all $u \in R, w \in B$ and profile P.

---

[3] A poignant variation of this observation is widely known as the Peter Principle.

A potential advantage of our Normalized Power Inequality measure is that it makes a first step towards a practically applicable measure[4] for power inequality. A fully individualized measure may provide a tool for actually "putting accusations of inequality to the test" in concrete individual cases, and not just pointing out "global unfairness" in society. Note, though, that our current measure falls short of providing such a test: one cannot compare just two individuals, since the Power Inequality effect only occurs *in expectation*, so to apply it one must obtain a representative sample of several individuals from each of the sets.

In some cases, Glass Ceiling proves to be a more sinister phenomenon than Power Inequality. An example of such a case is where the profile consists of a log of all the absence days of an employee (e.g., going on vacation or on leave, retiring early, etc). In this case, while we expect individuals that are less active to suffer the ill-effect of Power Inequality, the occurrence of a Full-Spectrum Glass Ceiling effect means that two individuals, one from R and another from B, with *identical* profiles, namely, who have taken vacations at the same time, do not have the same expected power. This may commonly be considered an unfair disadvantage.

Inequality Causes We focus on three possible causes of inequality. The first is *minority*, i.e., the fact that the number of women and men in the system is not necessarily equal. The second is *homophily*, a potential cause for inequality that was proposed and examined in [1], namely, the tendency of an individual to connect to similar individuals rather than non-similar individuals. One may conjecture that the tendency of men to connect to other men, and of women to connect to other women, particularly in combination with other causes, such as minority, may promote inequality. The third possible cause is a phenomenon called the *leaky pipeline*, whereby women quit work at a higher rate than men. It was proposed in [7] that this may be a key reason why women are underrepresented (specifically in the STEM fields, but in other domains as well).

Social Network Model Following [1], to facilitate the study of Power Inequality and Glass Ceiling in a concrete setting, we describe a mathematical model for social networks, where Power Inequality and Glass Ceiling naturally arise. Our model attempts to capture a collaborative social environment in which power is manifested by the number of connections made by an individual, and incorporates homophily, leaky pipeline and unequal initial conditions.

One major difference from the model of [1] is that while [1] considered the potential causes of minority and homophily, it did not consider the leaky pipeline as a possible cause. For this reason, we feel that the analysis of [1] may have failed to capture the phenomenon in its full affect.

Formally, we consider a network modeled as a multigraph, with two types of nodes, a set R of red nodes and a set B of blue nodes, representing our two populations. We would like to model a process by which individuals join their community at different times, stay in the community for a period of time during

---

[4] For example, a measure acceptable in the eyes of the law.

which they can be active, and then quit, or "retire". Formally, at every time t, a *class* t consisting of r new red nodes and b new blue nodes joins the network. At every time $\tau = t, t+1, \ldots, t+L-1$, every red (respectively, blue) node in class t decides either to be active in the network at time τ, with probability q_R (resp., q_B), or to be inactive, with probability $1 - q_R$ (resp., $1 - q_B$). At time $t + L$ all nodes in class t retire and become permanently inactive. Note that $q_R < q_B$ implies that red nodes tend to be inactive more often.

In this setting, power is represented by the ability to form connections, represented by links in the network. Link formation is modeled as follows. Every node starts with degree 0. At every time step, every active node v that is currently in the network attempts (simultaneously) to connect to a single other active node u in the network, chosen uniformly at random. If v and u are of the same color, a new edge (v, u) is added to the network. If v and u differ in color, however, then the nodes might fail to connect. Specifically, if $v \in B$ and $u \in R$, then with probability ρ_B the edge (v, u) is added, and with probability $1 - \rho_B$, the connection is not formed, and v resumes its search for a node to connect to. Conversely, if $v \in R$ and $u \in B$, then with probability ρ_R the edge (v, u) is added, and with probability $1 - \rho_R$, v resumes its search. Note that when $\rho_B < 1$ (respectively, $\rho_R < 1$), the network exhibits homophily by the blue (resp., red) nodes, i.e., nodes tend to connect to nodes of the same color.

The power of an individual v at time t is measured by its degree in the network at that time, i.e., $p(v, t) = d_t(v)$. We thus define the Power Inequality measure as

$$\mathsf{PI}(t) = \frac{\mathbb{E}\left[d_{t+L}(v)\right]}{\mathbb{E}\left[d_{t+L}(u)\right]}, \tag{3}$$

which is equal to the ratio between the expected degree of a red node v and a blue node u, both from class t, upon retirement. To compare individuals of similar profile, we represent the profile of an individual by a nonzero vector $\ell = (\ell_t, \ell_{t+1}, \ldots, \ell_{t+L-1}) \in \{0, 1\}^L$, specifying whether v was active (if $\ell_\tau = 1$) or inactive (if $\ell_\tau = 0$) at every time $t \leq \tau < t + L$. We can now provide a definition of the Normalized Power Inequality measure that is tailored to our setting, as

$$\mathsf{NPI}(t, \ell) = \frac{\mathbb{E}\left[d_{t+L}(v) \mid \mathcal{P}(v) = \ell\right]}{\mathbb{E}\left[d_{t+L}(u) \mid \mathcal{P}(u) = \ell\right]}, \tag{4}$$

which is equal to the ratio between the expected degree of a red node v with profile ℓ and a blue node u with profile ℓ, both from class t, upon retirement. Note that both PI and NPI do not depend on v (respectively, u) due to the symmetry between all the red (resp., blue) nodes in class t.

Contributions In this paper we study the phenomena of Power Inequality and Personalized Glass Ceiling discussed above, relate them to the causes of minority, homophily and leaky pipeline, and show how these three elements contribute to the creation (or absence) of the Power Inequality and Glass Ceiling effects.

The social network model just described allows us to conveniently represent the three causes discussed above. Let us define three ratios, the *minority ratio*, the *homophily ratio*, and the *leakage ratio*, as follows:

$$\mathcal{R}_{minority} = \frac{r}{b}, \qquad \mathcal{R}_{homophily} = \frac{\rho_B}{\rho_R} \cdot \frac{1 - \rho_R}{1 - \rho_B}, \qquad \mathcal{R}_{leakage} = \frac{q_R}{q_B}.$$

representing the level of imbalance between the two sub-populations in one of the three parameters discussed above. (The extreme cases where the denominator equals 0 require special treatment, but are in fact easier to analyze, and are expected to be rare.) In a completely balanced society, these ratios will all be equal to 1.

Our main result is that for a large enough L, a Full-Spectrum Glass Ceiling will gradually form for the red nodes, i.e., $\mathsf{NPI}(t, \ell) < 1$ for every profile ℓ, if

$$\mathcal{R}_{minority} \cdot \mathcal{R}_{homophily} \cdot \mathcal{R}_{leakage} < 1 .$$

Note that setting the parameters differently may create a Glass Ceiling for the blue nodes. Moreover, note that a bias in either one of the three ratios may "single-handedly" cause a Glass Ceiling effect, even if the other two ratios are both equal to 1, i.e., even if the society is balanced in two of the three parameters.

Let $\omega(\ell) = \sum_{i=t}^{t+L-1} \ell_i$ denote the *weight* of the profile ℓ, which can also be seen as the total amount of experience associated with the profile ℓ, and consider Eq. (4). We show that, in our model, $\mathbb{E}\left[d_{t+L}(u) \mid \mathcal{P}(u) = \ell\right]$ is linear[5] in $\omega(\ell)$, thus we have that $\mathbb{E}\left[d_{t+L}(v) \mid \mathcal{P}(v) = \ell\right] - \mathbb{E}\left[d_{t+L}(u) \mid \mathcal{P}(u) = \ell\right]$ is linear in the profile weight $\omega(\ell)$, implying that the difference in the power increases (linearly) with the experience of the individual.

We also derive the condition for the formation of a Power Inequality effect. This condition is technically more cumbersome. However, we observe that the Power Inequality measure $\mathsf{PI}(t)$ can be shown to be approximately equal to $\mathcal{R}_{leakage} \cdot \mathsf{NPI}(t, \ell)$. This implies that if $\mathcal{R}_{leakage} = 1$, then a Power Inequality will form for the red nodes, i.e., $\mathsf{PI}(t) < 1$, if

$$\mathcal{R}_{minority} \cdot \mathcal{R}_{homophily} < 1 .$$

In general, we show that for a large enough L, there exists a Strong Power Inequality for the red nodes if

$$q_R + \frac{r \cdot q_R \cdot (q_R - q_B \cdot \rho_R)}{r \cdot q_R + b \cdot q_B \cdot \rho_R} < q_B + \frac{b \cdot q_B \cdot (q_B - q_R \cdot \rho_B)}{b \cdot q_B + r \cdot q_R \cdot \rho_B} ,$$

and a Strong Power Inequality for the blue nodes if the inequality sign is flipped. Note that Power Inequality and Glass Ceiling are indeed two independent phenom-

---

[5]This is not necessarily true for all plausible models. For example, some natural models might assign more weight to vacations/leaves that take place early in the career.

ena. Namely, by setting $(r, b, q_R, q_B, \rho_R, \rho_B)$ to $(1, 4, 1, 0.5, 0.5, 0.5)$, we obtain a full-spectrum glass ceiling for the red nodes and a strong power inequality for the blue nodes.

2 The RLR Model

Our model incorporates the notion of a *leave*, where a node is temporarily inactive, and that of *retirement*, where a node becomes inactive forever, thus introducing the Recruitment, Leaves and Retirement (RLR) model. The system is described by a dynamically growing multigraph $G_t = (R \cup B, E_t)$ over two populations R and B, the sets of red and blue nodes. The red population $R = \bigcup_{t=1}^{\infty} R_t$ (respectively, $B = \bigcup_{t=1}^{\infty} B_t$) is a disjoint union of sets of size r (resp., b), i.e., $|R_t| = r$ and $|B_t| = b$ for all t. We call $C_t = R_t \cup B_t$ the *class* of year t. Note that $\{C_t\}_{t \geq 1}$ are pairwise disjoint, and that $|C_t| = r + b$.

For every node $v \in R_t$ (respectively, $v \in B_t$) in class t, we set v to be inactive at all times $i < t$ and $i \geq t + L$. For times $t \leq i < t + L$, we set v to be active with probability q_R (resp., q_B), and inactive otherwise. We denote the *activity profile* of node v from class t by $\mathcal{P}(v) = (\mathcal{P}_t(v), \mathcal{P}_{t+1}(v), \ldots, \mathcal{P}_{t+L-1}(v))$, where $\mathcal{P}_i(v) = 1$ if v is active at time i and 0 otherwise. The activity of a red (resp., blue) node v from class t is $\mathcal{P}_i(v) = 0$ for $i < t$, when v has not yet been recruited, as well as for $i \geq t+L$, when v has already retired. During the period $t \leq i < t+L$, v is employed, and $\mathcal{P}_i(v) = 1$ with probability q_R (resp., q_B) and 0 otherwise. We refer to L as the *career span*, and think of time t as the *time of recruitment*, the time $t + L$ as the *time of retirement*, and the times $t \leq i < t + L$ such that $\mathcal{P}_i(v) = 0$ as *leaves/vacations*. Another way of viewing this model is that at every time t, we add a class consisting of r red nodes and b blue nodes. This class stays in the work force for L time, where at every step, each red (respectively, blue) node in the class goes on a leave/vacation with probability $1 - q_R$ (resp., $1 - q_B$).

We now describe the process of edge formation. Starting with an empty graph $G_0 = (R \cup B, \emptyset)$, at every time t every active red (respectively, blue) node v chooses (simultaneously) an active node u, sampled uniformly at random. If v and u are of the same color, a new edge (v, u) is added to the multigraph, and if v and u differ in color, then with probability ρ_R (resp., ρ_B) the edge (v, u) is added, and with probability $1 - \rho_R$ (resp., $1 - \rho_B$), v resumes its search for a node to connect to.

Letting $d_t(v)$ denote the degree of node v at time t, we define Power Inequality and Normalized Power Inequality measures for the RLR model.

Power Inequality For nodes $v \in R_t$ and $u \in B_t$ in class t,

$$\mathsf{PI}(t) = \frac{\mathbb{E}\left[d_{t+L}(v)\right]}{\mathbb{E}\left[d_{t+L}(u)\right]} . \tag{5}$$

Normalized Power Inequality For nodes $v \in R_t$ and $u \in B_t$ in class t with the same profile $\ell = (\ell_t, \ell_{t+1}, \ldots, \ell_{t+L-1}) \in \{0, 1\}^L$

$$\text{NPI}(t, \ell) = \text{NPI}(v, u, t, \ell) = \frac{\mathbb{E}\left[d_{t+L}(v) \mid \mathcal{P}(v) = \ell\right]}{\mathbb{E}\left[d_{t+L}(u) \mid \mathcal{P}(u) = \ell\right]}. \tag{6}$$

Note that due to the symmetry between all the red (respectively, blue) nodes, these two definitions do not depend on the specific choices of v and u.

3 Analysis of the RLR Model

We now analyze the RLR model and state our main results. Denoting the profile of all the nodes by $\mathcal{P} = (\mathcal{P}(v) \mid v \in R \cup B)$, we prove the following theorem.

Theorem 1 *In the RLR model, for every red node $v \in R_t$ and $0 \leq j \leq L$, we have*

$$\mathbb{E}\left[d_{t+j}(v) \mid \mathcal{P}\right] = \sum_{i=t}^{t+j-1} \mathcal{P}_i(v) \cdot \left(1 + \frac{r_i}{r_i + b_i \cdot \rho_R} + \frac{b_i \cdot \rho_B}{b_i + r_i \cdot \rho_B}\right), \tag{7}$$

and symmetrically for every blue node $u \in B_t$, we have

$$\mathbb{E}\left[d_{t+j}(u) \mid \mathcal{P}\right] = \sum_{i=t}^{t+j-1} \mathcal{P}_i(u) \cdot \left(1 + \frac{b_i}{b_i + r_i \cdot \rho_B} + \frac{r_i \cdot \rho_R}{r_i + b_i \cdot \rho_R}\right), \tag{8}$$

where the expectation is conditioned on the activity profiles of the nodes.

Proof We prove Eq. (7). Consider a round i in which v is active. Then in this round v's degree will increase by 1 since v connects to another active node. In addition, each red (respectively, blue) node u chooses to connect to v with probability $p_{uv} = 1/(r_i + b_i \rho_R)$ (resp., $\rho_B/(b_i + r_i \rho_B)$). There are r_i (resp., b_i) such nodes, so

$$\mathbb{E}\left[d_{i+1}(v) - d_i(v) \mid \mathcal{P}\right] = \mathcal{P}_i(v)\left(1 + \frac{r_i}{r_i + b_i \cdot \rho_R} + \frac{b_i \cdot \rho_B}{b_i + r_i \cdot \rho_B}\right).$$

Summing over i from t to $t + j - 1$, Eq. (7) follows (and similarly for Eq. (8)). □

Theorem 2 (Conditional Expected Degree) *In the RLR model, for every class $t \geq L$, red node $v \in R_t$, profile ℓ, and $0 \leq j \leq L$, we have*

$$\mathbb{E}\left[d_{t+j}(v) \mid \mathcal{P}(v) = \ell\right]$$

$$= \left(1 + \frac{r \cdot q_R}{r \cdot q_R + b \cdot q_B \cdot \rho_R} + \frac{b \cdot q_B \cdot \rho_B}{b \cdot q_B + r \cdot q_R \rho_B} + o(1)\right) \sum_{i=t}^{t+j-1} \ell_i,$$

where $o(1) \to 0$ as $L \to \infty$. Similarly, for a blue node $u \in B_t$, we have

$$\mathbb{E}\left[d_{t+j}(u) \mid \mathcal{P}(u) = \ell\right]$$

$$= \left(1 + \frac{b \cdot q_B}{b \cdot q_B + r \cdot q_R \rho_B} + \frac{r \cdot q_R \cdot \rho_R}{r \cdot q_R + b \cdot q_B \rho_R} + o(1)\right) \sum_{i=t}^{t+j-1} \ell_i.$$

Consider the Glass Ceiling measure $\mathsf{NPI}(t, \ell)$ defined in Eq. (6). Note that $\mathsf{NPI}(t, \ell)$ does not depend on v and u and is given by the following direct corollary of Theorem 2.

Theorem 3 (Normalized Power Inequality)   *In the RLR model, for every career span L, class $t \geq L$, and profile ℓ, we have*

$$\mathsf{NPI}(t, \ell) = \frac{1 + \frac{r \cdot q_R}{r \cdot q_R + b \cdot q_B \cdot \rho_R} + \frac{b \cdot q_B \cdot \rho_B}{b \cdot q_B + r \cdot q_R \cdot \rho_B}}{1 + \frac{b \cdot q_B}{b \cdot q_B + r \cdot q_R \cdot \rho_B} + \frac{r \cdot q_R \cdot \rho_R}{r \cdot q_R + b \cdot q_B \cdot \rho_R}} + o(1),$$

where $o(1) \to 0$ as $L \to \infty$.

Corollary 1   *For a large enough career span $L > 0$ and a class $t \geq L$, a Full-Spectrum Glass Ceiling will form for the red nodes in class t if*

$$\mathcal{R}_{minority} \cdot \mathcal{R}_{homophily} \cdot \mathcal{R}_{leakage} < 1.$$

A Full-Spectrum Glass Ceiling will form for the blue nodes in class t if the inequality sign is flipped. Here we interpret $x/0 = \infty$ for every $x > 0$.

Proof This follows directly from Theorem 3 upon observing that for a large enough career span L, a Glass Ceiling will form for the red nodes if

$$\frac{(1 - \rho_R) \cdot r \cdot q_R}{r \cdot q_R + b \cdot q_B \cdot \rho_R} < \frac{(1 - \rho_B) \cdot b \cdot q_B}{b \cdot q_B + r \cdot q_R \cdot \rho_B},$$

and a glass will form for the blue nodes if the inequality sign is flipped. The corollary then follows upon noting that

$$\frac{(1 - \rho_R) \cdot r \cdot q_R}{r \cdot q_R + b \cdot q_B \cdot \rho_R} = 1 - (r \cdot q_R + b \cdot q_B) \cdot \frac{(r \cdot q_R + b \cdot q_B) \cdot \rho_B \cdot \rho_R + b \cdot q_B \cdot \rho_R (1 - \rho_B)}{(r \cdot q_R + b \cdot q_B \cdot \rho_R)(b \cdot q_B + r \cdot q_R \cdot \rho_B)}$$

$$\frac{(1 - \rho_B) \cdot b \cdot q_B}{b \cdot q_B + r \cdot q_R \cdot \rho_B} = 1 - (r \cdot q_R + b \cdot q_B) \cdot \frac{(r \cdot q_R + b \cdot q_B) \cdot \rho_B \cdot \rho_R + r \cdot q_R \cdot \rho_B (1 - \rho_R)}{(r \cdot q_R + b \cdot q_B \cdot \rho_R)(b \cdot q_B + r \cdot q_R \cdot \rho_B)}.$$

$\square$

The following are also direct corollaries of Theorem 2.

Corollary 2 *For every class* $t \geq L$, *and every node* v *in class* t, *we have that* $\mathbb{E}\left[d_{t+j}(v) \mid \mathcal{P}(v) = \ell\right]$ *is linear in the total amount of experience* $\omega(\ell) = \sum_{i=t}^{t+j-1} \ell_i$, *thus the difference* $\left|\mathbb{E}\left[d_{t+j}(v) \mid \mathcal{P}(v) = \ell\right] - \mathbb{E}\left[d_{t+j}(u) \mid \mathcal{P}(u) = \ell\right]\right|$ *increases (linearly) with the experience.*

Theorem 4 (Expected Degrees) *In the RLR model, for every class* $t \geq L$, *red node* $v \in R_t$ *and* $0 \leq j \leq L$, *we have*

$$\mathbb{E}\left[d_{t+j}(v)\right] = \left(1 + \frac{r \cdot q_R}{r \cdot q_R + b \cdot q_B \cdot \rho_R} + \frac{b \cdot q_B \cdot \rho_B}{b \cdot q_B + r \cdot q_R \cdot \rho_B} + o(1)\right) \cdot j \cdot q_R ,$$

where $o(1) \to 0$ *as* $L \to \infty$. *Similarly, for a blue node* $u \in B_t$, *we have*

$$\mathbb{E}\left[d_{t+j}(u)\right] = \left(1 + \frac{b \cdot q_B}{b \cdot q_B + r \cdot q_R \cdot \rho_B} + \frac{r \cdot q_R \cdot \rho_R}{r \cdot q_R + b \cdot q_B \cdot \rho_R} + o(1)\right) \cdot j \cdot q_B .$$

Next, consider the Power Inequality measure $\mathsf{PI}(t)$ defined in Eq. (5). Note that $\mathsf{PI}(t)$ does not depend on v and u and is given by the following direct corollary of Theorem 4.

Theorem 5 (Power Inequality) *In the RLR model, for every career span* L *and class* $t \geq L$, *we have*

$$\mathsf{PI}(t) = \frac{q_R}{q_B} \cdot \frac{\left(1 + \frac{r \cdot q_R}{r \cdot q_R + b \cdot q_B \cdot \rho_R} + \frac{b \cdot q_B \cdot \rho_B}{b \cdot q_B + r \cdot q_R \cdot \rho_B}\right)}{\left(1 + \frac{b \cdot q_B}{b \cdot q_B + r \cdot q_R \cdot \rho_B} + \frac{r \cdot q_R \cdot \rho_R}{r \cdot q_R + b \cdot q_B \cdot \rho_R}\right)} + o(1) ,$$

where $o(1) \to 0$ *as* $L \to \infty$.

The following is a direct corollary of Theorem 5.

Corollary 3 *For a large enough career span* L, *there exists a Strong Power Inequality for the red nodes if*

$$q_R + \frac{r \cdot q_R \cdot (q_R - q_B \cdot \rho_R)}{r \cdot q_R + b \cdot q_B \cdot \rho_R} < q_B + \frac{b \cdot q_B \cdot (q_B - q_R \cdot \rho_B)}{b \cdot q_B + r \cdot q_R \cdot \rho_B} .$$

A Strong Power Inequality for the blue nodes exists if the inequality sign is flipped.

We now prove Theorem 2. The proof uses the following lemma.

Lemma 1 *Fix nonnegative reals* $\rho, c_1, c_2 \geq 0$. *Let* $X \sim B(m_1, q_1)$ *and* $Y \sim B(m_2, q_2)$ *be two independent binomially distributed random variables. Denote*

$$f(X, Y) = \frac{c_1 + X}{c_1 + c_2 + X + \rho Y} \quad and \quad \gamma = \frac{\mathbb{E}[X]}{\mathbb{E}[X] + \rho \mathbb{E}[Y]} .$$

For $m_1, m_2 = \Theta(n)$, we have

$$|\mathbb{E}\,[f(X, Y)] - \gamma| \leq O(\sqrt{\ln n/n})\,.$$

Proof Letting $U_X = \mathbb{E}\,[X] + \alpha\sqrt{m_1}$ and $L_Y = \mathbb{E}\,[Y] - \alpha\sqrt{m_2}$, we consider the event $\mathcal{E} = \{X \leq U_X\} \cap \{Y \geq L_Y\}$. We have

$$\mathbb{E}\,[f(X, Y) - \gamma] = \Pr[\mathcal{E}] \cdot \mathbb{E}\,[f(X, Y) - \gamma \mid \mathcal{E}] + \Pr[\mathcal{E}^c] \cdot \mathbb{E}\,[f(X, Y) - \gamma \mid \mathcal{E}^c]\,, \quad (9)$$

where $\mathcal{E}^c$ is the complement of $\mathcal{E}$. Note that $f(X, Y)$ increases in X and decreases in Y, and that $0 \leq f(X, Y) \leq 1$ and $0 \leq \gamma \leq 1$. Plugging this into Eq. (9) yields

$$\mathbb{E}\,[f(X, Y) - \gamma] \leq 1 \cdot (f(U_X, L_Y) - \gamma) + \Pr[\mathcal{E}^c] \cdot 1\,.$$

Setting $\alpha = \sqrt{\ln n}$ and applying Hoeffding's inequality, we obtain

$$\Pr[\mathcal{E}^c] \leq \Pr[X > U_X] + \Pr[Y < L_Y] \leq 2e^{-2\alpha^2} = 2/n^2\,,$$

whereas

$$f(U_X, L_Y) - \gamma = \frac{c_1\mathbb{E}\,[Y] \cdot \rho - c_2\mathbb{E}\,[X] + \alpha \cdot (\sqrt{m_1} \cdot \mathbb{E}\,[Y] + \sqrt{m_2} \cdot \mathbb{E}\,[X])}{(\mathbb{E}\,[X] + \rho\mathbb{E}\,[Y])(c_1 + c_2 + U_X + \rho \cdot L_Y)}$$

$$\leq O(\sqrt{\ln n/n})\,,$$

thus proving $\mathbb{E}\,[f(X, Y) - \gamma] \leq O(\sqrt{\ln n/n})$ (and similarly for $\mathbb{E}\,[-f(X, Y) + \gamma]$).

$\square$

Corollary 4 *Following notations from Lemma 1, if the limits $\lim\limits_{n\to\infty} \frac{m_1}{n} = \beta_1$ and $\lim\limits_{n\to\infty} \frac{m_2}{n} = \beta_2$ exist, we get*

$$\mathbb{E}\,[f(X, Y)] = \frac{\beta_1 q_1}{\beta_1 q_1 + \rho \cdot \beta_2 q_2} + o(1)\,,$$

where $o(1)$ tends to 0 as $n \to \infty$.

Proof of Theorem 2. We prove the claim for the red nodes (a similar proof holds for the blue nodes). By Eq. (7), for every red node v we have

$$\mathbb{E}\,[d_{t+L}(v) \mid \mathcal{P}(v) = \ell] = \sum_{i=t}^{t+L-1} \ell_i \cdot \mathbb{E}\left[1 + \frac{r_i}{r_i + b_i \cdot \rho_R} + \frac{b_i \cdot \rho_B}{b_i + r_i \cdot \rho_B}\right]$$

Whenever $\ell_i = 0$, the summand is equal to 0. When $\ell_i = 1$, since the distribution of r_i (respectively, b_i) is independent of i, the summand is also independent of i. Hence

$$\mathbb{E}\left[d_{t+L}(v) \mid \mathcal{P}(v) = \ell\right]$$

$$= \left(\sum_{i=t}^{t+L-1} \ell_i\right) \cdot \mathbb{E}\left[1 + \frac{1+\hat{r}_t}{(1+\hat{r}_t) + b_t \cdot \rho_R} + \frac{b_t \cdot \rho_B}{b_t + (1+\hat{r}_t) \cdot \rho_B}\right],$$

where $\hat{r}_t = \sum_{w \in R \setminus \{v\}} \mathcal{P}_t(w)$ is the number of red nodes active at time t not including v. Note that at time $t \geq L$, there are potentially $L \cdot r$ active nodes, thus $\hat{r}_t \sim B(L \cdot r - 1, q_R)$ is a binomially distributed random variable. Applying Corollary 4 twice, once with (X, Y, ρ, c_1, c_2) set to $(\hat{r}_t, b_t, \rho_R, 1, 0)$ and once with (X, Y, ρ, c_1, c_2) set to $(b_t, \hat{r}_t, \rho_B, 0, \rho_B)$, the claim follows. □

4 Discussion

Let us conclude with a discussion of the meaning and implications of the last result. Consider, as a concrete example of relevance, a work environment with men and women. We model the leaky pipeline phenomena where women leave the network at a higher rate than men as $q_R < q_B$. While in many fields men and women arrive to the network at similar rates (i.e., $r = b$), a Glass Ceiling can also be generated if women arrive at a higher rate than men, i.e. $r > b$, as long as their leaving rate is high enough, they are discriminated enough, or have enough low self-homophily, namely $r \cdot q_R \cdot \rho_B \cdot (1 - \rho_R) < b \cdot q_B \cdot \rho_R \cdot (1 - \rho_B)$.

Corollary 3 formalizes the connection between the following three elements, which contribute to the formation and destruction of Glass Ceilings: initial majority, given by the relation between r and b; the leaving rates for the leaky pipeline, modeled as q_R and q_B; and the rate of homophily, given by ρ_R and ρ_B.

If no homophily exists (i.e., $\rho_R = \rho_B = 1$) then by symmetry, nodes with identical profiles will have identical degree distribution, thus there will be no Glass Ceiling. Dually, in the case of total homophily (i.e., $\rho_R = \rho_B = 0$), two disjoint networks will be created, one with only blue nodes and the other with only red nodes. By symmetry, in each network and each time step, each active node will increase its expected degree by 2, thus all nodes with the same profile will have the same expected degree.

If partial homophily exists (i.e., $0 < \rho_R, \rho_B < 1$), then a Glass Ceiling will be generated for the current minority, scaled by the homophily parameters. Specifically, a Glass Ceiling for R will form if the product of the ratios $\mathcal{R}_{minority}$, $\mathcal{R}_{homophily}$ and $\mathcal{R}_{leakage}$ is smaller than 1. It is important to consider all three elements, since if all three ratios are nontrivial (i.e., equal to neither 0 nor infinity), then any one of them can overthrow the other two by solely generating or breaking the Glass Ceiling.

Finally, let us mention that another attempt to formalize a Glass Ceiling effect can be found in [4], where four criteria are formalized for the generation of Glass

Ceiling. This work, however, is very different from ours for a few reasons. First, [4] does not take into account the personal capabilities and experience of each individual. Second, their work attempts to quantify the Glass Ceiling by observing empirical data, whereas we attempt to model a social network and explain the causes for (and quantify) the Glass Ceiling.

Acknowledgements Supported in part by the Israel Science Foundation (grant 1549/13).

References

1. Avin, C., Keller, B., Lotker, Z., Mathieu, C., Peleg, D., Pignolet, Y.-A.: Homophily and the glass ceiling effect in social networks. In: Proceedings of the ACM Conf. on Innovations in Theoretical Computer Science, pp. 41–50 (2015)
2. Ceci, S.J., Williams, W.M.: Sex differences in math-intensive fields. Curr. Dir. Psychol. Sci. **19**, 275–279 (2010). doi:10.1177/0963721410383241
3. Ceci, S.J., Williams, W.M., Barnett, S.M.: Women's underrepresentation in science: sociocultural and biological considerations. Psychol. Bull. **135**(2), 218 (2009)
4. Cotter, D.A., Hermsen, J.M., Ovadia, S., Vanneman, R.: The glass ceiling effect. Soc. Forces **80**(2), 655–681 (2001)
5. Diekman, A.B., Brown, E.R., Johnston, A.M., Clark, E.K.: Seeking congruity between goals and roles a new look at why women opt out of science, technology, engineering, and mathematics careers. Psychol. Sci. **21**(8), 1051–1057 (2010)
6. Federal Glass Ceiling Commission: Solid Investments: Making Full Use of the Nation's Human Capital, p. 4 (1995)
7. Griffith, A.L.: Persistence of women and minorities in stem field majors: is it the school that matters? Econ. Educ. Rev. **29**(6), 911–922 (2010)
8. Gürer, D., Camp, T.: Investigating the incredible shrinking pipeline for women in computer science. Final report–NSF project, 9812016 (2001)

Elites in Social Networks: An Axiomatic Approach

Chen Avin, Zvi Lotker, David Peleg, Yvonne-Anne Pignolet, and Itzik Turkel

Abstract Recent evidence shows that in many societies the relative sizes of the economic and social *elites* are continuously shrinking. Is this a *natural* social phenomenon? We try to address this question by studying a special case of a core-periphery structure composed of a social *elite*, namely, a relatively small but well-connected and highly influential group of powerful individuals, and the rest of society, the *periphery*. Herein, we present a novel axiom-based model for the mutual influence between the elite and the periphery. Assuming a simple set of axioms, capturing the elite's *dominance*, *robustness* and *compactness*, we are able to draw strong conclusions about the elite-periphery structure. In particular, we show that the elite size is *sublinear* in the network size in social networks adhering to the axioms. We note that this is in controversy to the common belief that the elite size converges to a linear fraction of society (most recently claimed to be 1%).

Keywords Social networks • Core-periphery • Elite • Block model • Partition • Axioms • Influence • Social structure

1 Introduction

In his book *Mind and Society* [21], Vilfredo Pareto wrote what is by now widely accepted by sociologists: "Every people is governed by an *elite*, by a chosen element of the population". Indeed, with the exception of some rare examples of utopian or totally egalitarian societies, almost all societies exhibit an (often radically) uneven distribution of power, influence, and wealth among their members,

C. Avin (✉) • Z. Lotker • I. Turkel
Ben Gurion University of the Negev, Be'er Sheva, Israel
e-mail: avin@cse.bgu.ac.il; zvilo@cse.bgu.ac.il; turkel@cse.bgu.ac.il

D. Peleg
Weizmann Institute of Science, Rehovot, Israel
e-mail: david.peleg@weizmann.ac.il

Y.-A. Pignolet
ABB Corporate Research, Baden, Switzerland
e-mail: yvonne-anne.pignolet@ch.abb.com

© Springer International Publishing AG 2017
E. Shmueli et al. (eds.), *3rd International Winter School and Conference on Network Science*, Springer Proceedings in Complexity, DOI 10.1007/978-3-319-55471-6_7

and, in particular, between the *elite* and its complement, sometimes referred to as the *masses*. Typically, the elite is small, powerful and influential, whereas the complementary part of society is larger, less organized, and less dominant.

Looking more closely at social networks, the distinction between the elite and the rest of society can be viewed as a special case of a more general division that occurs in most complex networks, usually referred to as a *core-periphery* partition of the network [5]. The core-periphery structure is arguably the most high-level structure of society, and the problem of identifying this partition and understanding its basic properties has recently received increasing attention [18, 23, 26]. Generally speaking, the core vertices are more highly connected and feature higher centrality values than the periphery vertices; these properties are naturally shared by elites in social networks.

However, we argue that elites have some additional significant properties, which distinguish them as a special class of cores worthy of independent study. In particular, these properties imply two notable characteristics of elites, namely, that they are relatively *small* and that they possess a *disproportionate* fraction of the power, resources, and influence in society.

In this paper we study the properties of social elites. Our main contribution is a characterization of elites, i.e., a set of properties (formulated as "axioms") concerning influence and density that any elite must possess. We stress that we do not claim these axioms hold for every core-periphery partition, nor do we claim that every social network admits a core-periphery partition that satisfies the axioms; in fact, it is easy to find examples of both real-life complex networks and classical evolutionary network models in which our axioms are not met by any core-periphery partition. Rather, we focus on the class of social networks that *do* admit core-periphery partitions that satisfy our axioms, referred to hereafter as the class of *elite-centered* social networks.

A small illustrative example of the terms we use is provided in Fig. 1. It presents the network of the top 139 Marvel [1] superheroes and the 924 links interconnecting them, partitioned into an elite and a periphery as shown by different vertex colors. Two striking features can be clearly observed in this figure. First, the elite (containing, e.g., Captain America, Spiderman, and Thor), depicted in Fig. 1b, is dense and organized, while the periphery, presented in Fig. 1c, is much sparser and less structured. Second, the size of the core is "only" 27 vertices, with 112 vertices in the periphery. Note that despite this considerable size difference, the elite and the periphery have almost the same number of internal edges (≈ 250).

Our axiomatic characterization does not lead to pinpointing a single definition for the elite in a given social network. However, it is powerful enough to allow us to derive several conclusions concerning basic properties of the elite in society. Our main conclusion applies to the *size* of the elite. Recent reports show that the gap between the richest people and the masses keeps increasing, and that decreasingly fewer people amass more and more wealth [13, 20]. The question raised by us is: can society help it, or is this phenomenon an unavoidable by-product of some inherent natural properties of society? We claim that in fact, one can predict the shrinkage of elite size over time (as a fraction of the entire society size) based on the very

Fig. 1 Fictional illustrative example: the social network of the Marvel's superheroes. Two heroes are linked if they appeared together in many comic book titles [1]. (**a**) The network (139 superheroes, 924 edges), partitioned into an elite (*red vertices* and *internal edges*) and a periphery (*green vertices* and *internal edges*). *Blue "crossing" edges* connect elite and periphery vertices. (**b**) The (dense) elite subgraph (27 vertices, 252 edges). (**c**) The (sparser) periphery subgraph (112 vertices, 249 edges)

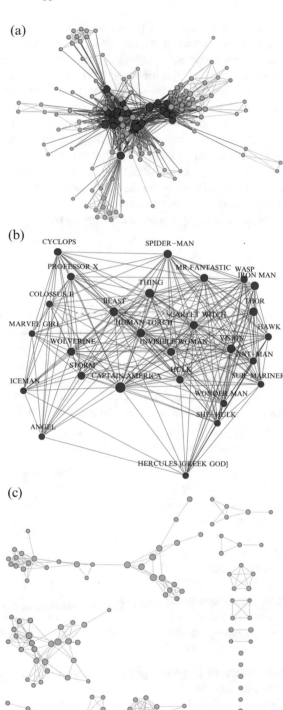

nature of social elites. In particular, in our model, such shrinkage is the natural result of a combination of two facts: First, *society grows*, and second, *elites are denser* than peripheries (informally, they are much better connected). Combining these facts implies that the fraction of the total population size comprising dense elites will decrease as the population grows with time. We prove this formally in Theorem 2.

A dual question we are interested in concerns the stable size of the elite in a growing society: How small can the elite be while still maintaining its inherent properties? In general, elites of constant size can exist in societies where influence might be sharply asymmetric. In contrast, we show that if the social network is unweighted and undirected then an elite cannot be smaller than $\Omega(\sqrt{m})$, where m is the number of network edges.

Supporting empirical results are presented in [4].

2   An Axiomatic Approach

The common approach to explaining empirical results on social networks is based on providing a new concrete (usually random) *evolutionary model* and comparing its predictions to the observed data. In contrast, we follow an *axiomatic approach* to the questions at hand. This approach is based on postulating a small set of axioms, capturing certain expectations about the network structure and the basic properties that an elite must exhibit in order to maintain its power in the society. Our axioms are inspired by elite definitions like the one from Wikipedia, by which:

> In political and sociological theory, an elite is a small group of people who control a disproportionate amount of wealth or political power.

To conceptualize these informal definitions, we employ the fundamental notion of *influence* among groups of vertices, and propose three independent properties related to the influence between the elite and the periphery. The underlying assumption is that the excess influence of the elite allows it on the one hand to control the rest of the population, and on the other to protect its members from being controlled by others outside the elite. We refer to these two properties as *dominance* and *robustness* respectively. In addition, the "wealth" is shared by the elite few, implying that on average, the elite members hold much more influence than individuals in the periphery. We refer to this property as *compactness*. We characterize *elite-centered* social networks as the class of social networks that admit core-periphery partitions satisfying these three properties. Next we make these properties more formal.

Influence and Core-Periphery Partition   We consider *influence* to be a measurable quantity between any two *groups* of people from the population, X and Y, denoted by $\mathcal{I}(X, Y)$. The groups X and Y are not necessarily distinct, and we are also interested in the *internal influence* exerted by the vertices of a group X on themselves, referred to by $\mathcal{I}(X, X)$. We provide the abstract notion of influence in social networks a concrete interpretation based on edge weights. We also assume that every individual has a self-opinion (modeled as a *self-loop.*) We denote the set of core vertices by $\mathcal{C}$ and the rest of society (the periphery) by $\mathcal{P}$. We call the pair $(\mathcal{C}, \mathcal{P})$, which satisfies $\mathcal{C} \cap \mathcal{P} = \emptyset$ and $\mathcal{C} \cup \mathcal{P} = V$, a *core-periphery* partition, and study the four influence quantities $\mathcal{I}(\mathcal{C}, \mathcal{C}), \mathcal{I}(\mathcal{P}, \mathcal{P}), \mathcal{I}(\mathcal{C}, \mathcal{P})$ and $\mathcal{I}(\mathcal{P}, \mathcal{C})$.

Formally, we model a social network as a directed, weighted graph $G = (V, E, \omega)$, with a set V of n vertices representing the members of society, connected by a set $E \subseteq V \times V$ of m directed edges, and a positive weight function $\omega : E \to \mathbb{R}$ such that $\omega(e) > 0$ for every $e \in E$. We are interested in the *relative* (and not absolute) influence between vertices, so we shall initially fix the weights of self loops to 1, thus defining a "unit of influence", and assume that all other weights are relativized to that unit, and next *normalize* the weight function ω so that $\sum_{e \in E} \omega(e) = |E| = m$. For a set of edges $E' \subseteq E$, define the weight of E' as $\omega(E') = \sum_{e \in E'} \omega(e)$. Given an undirected network, we consider each undirected edge as two equal weight directed edges. Given an unweighted network, we consider all edges to have weight one.

For every vertex v and set of vertices X, let the set of directed edges connecting v to vertices in X be denoted by $E(v, X)$. Similarly, for vertex sets $X, Y \subseteq V$, let $E(X, Y)$ denote the set of directed edges connecting vertices in X to vertices in Y. Based on the edge weights, we define the influence of X on Y, for $X, Y \subseteq V$, as

$$\mathcal{I}(X, Y) = \omega(E(X, Y)). \tag{1}$$

Note that in general $\mathcal{I}(X, Y) \neq \mathcal{I}(Y, X)$. However, if the social network is undirected then $\mathcal{I}(X, Y) = \mathcal{I}(Y, X)$ for every $X, Y \subseteq V$. In addition we define the *total power* of a set X to be

$$\mathcal{I}(X) = \mathcal{I}(X, X) + \mathcal{I}(X, V \setminus X). \tag{2}$$

Given a core-periphery partition $(\mathcal{C}, \mathcal{P})$ of V, the edge set E can be partitioned into four disjoint edge sets $E(\mathcal{C}, \mathcal{C}), E(\mathcal{C}, \mathcal{P}), E(\mathcal{P}, \mathcal{C})$ and $E(\mathcal{P}, \mathcal{P})$. Looking at the *adjacency matrix* $A(G)$ of the core-periphery network G [7], these sets correspond to the four basic parts of the *block-model representation* [14] of $A(G)$.

Elite-Periphery Axioms We now propose three simple axioms that capture what we consider to be basic structural properties required of the core-periphery partition $(\mathcal{E}, \mathcal{P})$ in *elite-centered* social networks,[1] namely, *dominance*, *robustness*, and *compactness*. To state our axioms we first define three corresponding quantitative measures for the dominance, robustness, and compactness of the elite $\mathcal{E}$ in a given $(\mathcal{E}, \mathcal{P})$ partition.

---

[1]To emphasize our focus on networks whose core is an elite, we denote the core set of the partition by $\mathcal{E}$ rather than $\mathcal{C}$.

1. **Dominance:** The first measure, referred to as the *elite dominance*, concerns the balance of forces exerted on the *periphery*; namely, it compares the influence of the elite on the periphery with the internal influence of the periphery. Formally,

$$\text{dom}(\mathcal{E}) = \mathcal{I}(\mathcal{E}, \mathcal{P})/\mathcal{I}(\mathcal{P}, \mathcal{P}),$$

and the first axiom is:

(A1) Elite-Dominance: dom $(\mathcal{E}) \geq c_d$, for a fixed constant $c_d > 0$

This axiom states that the elite *dominates* the rest of society, namely, that the *external* influence maintained by the elite $\mathcal{E}$ on the periphery $\mathcal{P}$ is higher (or at least not significantly lower) than the *internal* influence that the periphery has on itself. Such high dominance is essential for the elite to be able to maintain its superior status in society.

2. **Robustness:** The second measure, referred to as the *elite robustness*, concerns the forces exerted on the *elite*, namely, it compares the internal influence of the elite with the influence of the periphery on the elite. Formally,

$$\text{rob}(\mathcal{E}) = \mathcal{I}(\mathcal{E}, \mathcal{E})/\mathcal{I}(\mathcal{P}, \mathcal{E}),$$

and the second axiom is:

(A2) Elite-Robustness: $\text{rob}(\mathcal{E}) \geq c_r$, for a fixed constant $c_r > 0$

This axiom claims that elite is *robust*; to maintain its cohesiveness and be able to stick to its opinions, the elite must be able to resist "outside" pressure in the form of the periphery's external influence. To achieve that, the *internal* influence of the elite $\mathcal{E}$ on itself must be greater (or at least not significantly less) than the *external* influence exerted on $\mathcal{E}$ by the periphery.

3. **Compactness:** The third measure, referred to as the *elite compactness*, concerns the *disproportionality* between the elite's *power* and *size*. Let $\delta_X = \frac{\log \mathcal{I}(X)}{\log |X|}$ denote the *log-density* of a set $X \subseteq V$. Then

$$\text{comp}(\mathcal{E}) = \delta_{\mathcal{E}}/\delta_V,$$

and the third axiom is:

(A3) Elite-Compactness: $\text{comp}(\mathcal{E}) \geq 1 + c_c$, for a fixed constant $c_c > 0$

This axiom states that the elite members are more compact (or dense) than the entire network. This means that on average an elite member holds significantly more power than an arbitrary member of society.

The three axioms are illustrated graphically in Fig. 2. We say that a family of n-vertex networks G_n, for growing n, satisfies the axiom A if there exists some n_0 such that G_n satisfies A for every $n \geq n_0$.

Before showing the implications of these axioms we show that the three axioms are independent. For any two axioms out of the three, there exist a social network and a core-periphery partition that satisfies the two axioms but not the third. More formally, we have the following.

Theorem 1 (Axiom Independence) *Axioms (A1), (A2), (A3) are independent, namely, assuming any two of them does not imply the third.*

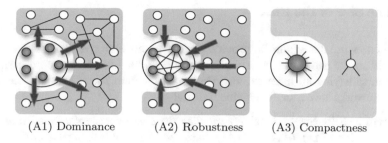

Fig. 2 Graphical illustration of the three axioms. Elite vertices are *gray*. (**A1**) The elite's external influence (*blue edges*), $\mathcal{I}(\mathcal{E}, \mathcal{P})$, dominates the periphery's internal influence (*black edges*), $\mathcal{I}(\mathcal{P}, \mathcal{P})$. (**A2**) The internal influence of the elite, $\mathcal{I}(\mathcal{E}, \mathcal{E})$, is robust to the periphery's external influence, $\mathcal{I}(\mathcal{P}, \mathcal{E})$. (**A3**) The elite is more compact and its average individual is more powerful than an average individual in the society

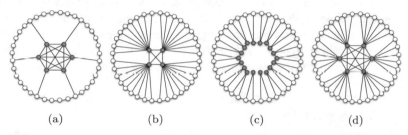

Fig. 3 Network examples demonstrating the independence of the axioms (the *gray* vertices form the core). (**a**) The core is robust and compact but not dominant. (**b**) The core is dominant and compact but not robust. (**c**) The core is dominant and robust but not compact. (**d**) An example of a network satisfying all three axioms

Proof Sketch. We prove the theorem by considering three examples of families of n-vertex (undirected, unweighted) networks and core-periphery partitions for them, described next. Each of these partitions satisfies two of the axioms, but violates the third, implying axiom independence. The first network and partition (Fig. 3a), depict a core that is robust and compact but whose dominance tends to zero as the network size n grows to infinity. The second example (Fig. 3b) describes a core that is dominant and compact but whose robustness tends to zero as the network grows. The last example (Fig. 3c) describes a core that is dominant and robust its compactness dominance tends to one as the network size n grows to infinity. i.e., the average degree of core members and periphery members is almost the same. □

We observe that, as one can easily verify, there are certain networks for which no core-periphery partition satisfies all three axioms (A1), (A2), and (A3) simultaneously. Interestingly, in the special case of undirected networks, axioms (A1) and (A2) are "inversely dependent", namely, every network and every core-periphery partition must satisfy at least one of them. This implies that there are no unweighted networks that disobey all three axioms.

3   The Size of the Elite

We can now use our axioms to provide bounds on the elite size. The class of *elite-centered social networks* consists of social networks that admit a core-periphery partition satisfying all three axioms. Our main theorem shows that elite-centered social networks have a sublinear elite. Formally,

Theorem 2 (Elite Size)   *If* $(\mathcal{E}, \mathcal{P})$ *satisfies the dominance, robustness, and compactness axioms (A1), (A2), and (A3), then the elite size is* sublinear *in the size of society, namely,*

$$c \cdot n^{\frac{\delta_V}{\delta_{\mathcal{E}}}} \leq |\mathcal{E}| \leq n^{\frac{1}{1+c_c}} .$$

We find it remarkable that three simple and intuitive assumptions lead to such a strong implication on the elite size. Note that Theorem 2 is controversial to the common belief that the elite size converges to a linear fraction of a society's size (most recently claimed to be 1% [22]). This discrepancy may perhaps be attributed to the fact that our axiom-based approach characterizes the elite differently than in previous approaches.

To prove Theorem 2 we first observe that every network satisfying the axioms has the following properties.

Lemma 1   *If* $(\mathcal{E}, \mathcal{P})$ *satisfies the dominance and robustness axioms (A1) and (A2), then the total influence of the elite is at least a fraction of the total influence in society, namely, for some constant* $c_1 > 0$,

$$c_1 \cdot m \leq \mathcal{I}(\mathcal{E}) \leq m.$$

Proof We first note that since $\omega(E) = m$ we have:

$$\mathcal{I}(\mathcal{E}, \mathcal{E}) + \mathcal{I}(\mathcal{E}, \mathcal{P}) + \mathcal{I}(\mathcal{P}, \mathcal{E}) + \mathcal{I}(\mathcal{P}, \mathcal{P}) = m, \tag{3}$$

so $\mathcal{I}(\mathcal{E}) \leq m$. By Eq. (3) and Axioms (A1) and (A2),

$$\mathcal{I}(\mathcal{E}, \mathcal{E}) + \mathcal{I}(\mathcal{E}, \mathcal{P}) + \frac{\mathcal{I}(\mathcal{E}, \mathcal{E})}{c_r} + \frac{\mathcal{I}(\mathcal{E}, \mathcal{P})}{c_d} \geq m,$$

so

$$\left(1 + \frac{1}{\min(c_r, c_d)}\right) \mathcal{I}(\mathcal{E}) \geq m,$$

hence $\mathcal{I}(\mathcal{E}) \geq c_1 \cdot m$ for constant $c_1 = (1 + \frac{1}{\min(c_r, c_d)})^{-1}$. $\square$

Using Axiom (A3), the elite size can now be tightly bounded in terms of the compactness parameters $\delta_{\mathcal{E}}$ and δ_V, establishing Theorem 2.

Proof (of Theorem 2) Recalling that $m = n^{\delta_V}$ and $\mathcal{I}(\mathcal{E}) = |\mathcal{E}|^{\delta_\mathcal{E}}$, the proof follows directly from Lemma 1 and Axiom (A3), which states that $\delta_V/\delta_\mathcal{E} \leq 1/(1 + c_c)$. □

In reality, the question of an upper bound for the "typical" elite is unanswered: does the "universal" size of elites (if it exists) converge to a linear, or a sublinear, function of the network size? In [4] we present evidence that many social networks are elite-centered (namely, satisfy our axioms), which indicates sublinear elites.

Interestingly, if Axiom (A3) does not hold, then it is possible for the elite to be of linear size. This will be the case, for example, in social networks where Dunbar's theory holds [12]. Dunbar suggested a cognitive limit to the number of people with whom one can maintain stable social relationships. If this is the case, then one can show that an elite that satisfies Axioms (A1) and (A2) *must* be of linear size. One can even claim a slightly stronger result, stating that the elite's average degree is bounded from above by a constant times the average degree in the network (which necessitates linear elite size). Formally, we state the following.

Lemma 2 *If $(\mathcal{E}, \mathcal{P})$ satisfies the dominance and robustness axioms (A1) and (A2), but has, for some constant c,*

$$\mathcal{I}(\mathcal{E})/|\mathcal{E}| \leq c \cdot \mathcal{I}(V)/|V|,$$

then $|\mathcal{E}| \geq c_3 \cdot n$ for some constant c_3.

Proof By Lemma 1, and since here $\mathcal{I}(\mathcal{E}) \leq |\mathcal{E}|cm/n$, we have $|\mathcal{E}| \geq c_1 n/c$. □

We now turn lower bounds on the elite size. How small can the elite be while still maintaining its power and satisfying the axioms? Let us first observe that in the general case of weighted or directed networks (such as twitter for example), no nontrivial lower bounds hold, and the network may have an extremely small elite (possibly even constant size) that satisfies our axioms.

Lemma 3 *There are (directed or weighted) networks for which an $(\mathcal{E}, \mathcal{P})$ partition satisfies the dominance, robustness, and compactness axioms (A1), (A2), and (A3), and the elite has a constant number of members.*

Proof For directed networks, a classical example is the *star* graph, where a single vertex (the center, forming the elite) has a directed edge to each of the periphery vertices (with no incoming edges). Clearly the star center dominates the periphery, and it is robust and compact.

Next consider undirected weighted graphs. Consider a tree network with $n = 2k$ vertices, so $m = 4k - 1$ (including self-loops). The weight of each self-loop is 1, totaling $2k$. Now the tree is constructed from two stars, with uniform edge weights of $1/2$, plus an edge of weight k connecting the two centers of the stars. It is easy to check that the sum of the edge weights is m. The elite consisting of the two star centers satisfies all three axioms. □

In contrast, in undirected unweighted networks, the following lower bound can be established for elite size.

Theorem 3 *In an unweighted and undirected network G with a core-periphery partition* $(\mathcal{E}, \mathcal{P})$, *if the core* $\mathcal{E}$ *satisfies the dominance and robustness axioms (A1) and (A2), then its size satisfies* $|\mathcal{E}| \geq c_4 \cdot \sqrt{m}$ *for some constant* $c_4 > 0$.

Proof In the undirected case $\mathcal{I}(\mathcal{E}, \mathcal{P}) = \mathcal{I}(\mathcal{P}, \mathcal{E})$, so

$$\mathcal{I}(\mathcal{E}, \mathcal{E}) + \mathcal{I}(\mathcal{E}, \mathcal{P}) + \mathcal{I}(\mathcal{P}, \mathcal{P}) = m. \tag{4}$$

By the two axioms and since G is undirected, we have

$$\mathcal{I}(\mathcal{E}, \mathcal{E}) \geq c_r \cdot \mathcal{I}(\mathcal{P}, \mathcal{E}) \geq c_r c_d \cdot \mathcal{I}(\mathcal{P}, \mathcal{P}).$$

Combining this with Eq. (4), we obtain

$$m \leq \left(1 + \frac{1}{c_r} + \frac{1}{c_r c_d} \right) \mathcal{I}(\mathcal{E}, \mathcal{E}).$$

Hence, when setting $c_2 = (1 + 1/c_r + 1/(c_r c_D))^{-1}$ it holds that

$$\mathcal{I}(\mathcal{E}, \mathcal{E}) \geq c_2 \cdot m. \tag{5}$$

Graph-theoretical considerations dictate that $\mathcal{I}(\mathcal{E}, \mathcal{E}) \leq \binom{|\mathcal{E}|}{2} \leq |\mathcal{E}|^2$, implying that $|\mathcal{E}| \geq \sqrt{\mathcal{I}(\mathcal{E}, \mathcal{E})}$. Combined with Eq. (5), the theorem follows. $\square$

One can also show an example of what we call a *purely elitistic society*, where the elite reaches its minimum possible size of $\Theta(\sqrt{m})$ in undirected, unweighted networks. See Fig. 3d for an illustrative example.

We remark that in addition to the theoretical results on elite axioms and properties, we also studied some real networks, in order to examine the extent to which our axioms manifest in reality, and provide evidence for the existence of real elite-centered social networks [4].

4 Related Work

The axiomatic approach has been used successfully in many fields of science, such as mathematics, physics, economy, sociology and computer science. See [2, 16] for two examples in areas related to ours.

A variety of notions for measuring influence in a network and for core-periphery partitions have been developed in the past (see the recent survey [9]). Borgatti and Everett [5] measure the similarity between the adjacency matrix of a graph and the block matrix $\left(\begin{smallmatrix} 1 & 1 \\ 1 & 0 \end{smallmatrix} \right)$. This captures the intuition that social networks have a dense, cohesive core and a sparse, disconnected periphery. an intuition also reflected in the axioms postulated herein.

Methods for identifying core-periphery structures and partitioning networks include algorithms for detecting (along with statistical tests for verifying) a-priori hypotheses [6], a coefficient measuring if a network exhibits a core-periphery dichotomy [18], a method for extracting cores based on a modularity parameter [11], a centrality measure computed as a continuous value along a core-periphery spectrum [23], a coreness value attributed to each vertex, qualifying its position and role based on random walks [24], a detection method using spectral analysis and geodesic paths [10], and a decomposition method using statistical inference [26]. The recent [25] argues that the core-periphery structure is simply the result of several overlapping communities and proposes a community detection method coping with overlap. None of these works consider the asymptotic size of a core/elite and the possibility that its size is sublinear in the population size.

One of the first papers to focus on the fact that the highest degree vertices are well-connected [27] coined the term *rich-club coefficient* for the density of the vertices of degree k or more. Mislove et al. [19] defined the *core* of a network to be any (minimal) set of vertices that satisfies two properties. First, the core must be essential for ensuring the connectivity of the network (i.e., removing it breaks the remaining vertices into many small, disconnected clusters). Second, the core must be strongly connected with a relatively small diameter. Mislove et al. used an approximation technique based on removing increasing numbers of the highest degree vertices (rich clubs) and analyzing the connectivity of the remaining graph. The graphs studied in [19] have a densely connected core comprising of between 1% and 10% of the highest degree vertices, such that removing this core completely disconnects the graph. Thus, the authors provide further evidence that rich clubs are crucial in social networks and satisfy their core properties.

A very different perspective is offered in [17]; a network formation game is studied, where benefits from connections exhibit decreasing returns and decay with network distance. In line with our axioms, the equilibria of this game form core-periphery structures. Another network formation game is developed in [15], where players invest in information acquisition. The authors show what they call "The Law of the Few": the economic forces are leading to a robust equilibrium where the majority of individuals to obtain most of the information from a very small subset of the group. The size of this subset is sublinear, so its fraction out of the population converges to zero. While these results hold under a more specific set of assumptions, they confirm the results derived from our more general axioms.

Recently, [3] used ideas presented in this paper to study the influence properties of the set of *founders*, the vertices arriving first, in the preferential attachment model of [8] under different model parameters. If the number of edges in the model is linear in the number of vertices (i.e., edge and vertex events happen with constant probability), then networks generated by preferential attachment must have a *linear* size founders set to be dominant, implying that this set will not satisfy our third axiom. On the other hand, if the number of edges in the model is super-linear in the number of vertices (i.e., the probability of vertex events decreases to zero over time), then the generated networks feature a *sublinear* size founders set that is dominant. This also demonstrates that both linear and sublinear cores are possible, depending on the network type.

5   Conclusion

In this article, we provide axioms modeling the influence relationships between the
elite and the periphery. We prove that for a core-periphery partition that satisfies our
axioms, the core forms an elite of sublinear size in the number of network vertices.
In particular, this means that an elite is much smaller than a constant fraction of the
network, evidence of which is often observed in the widening gap between the very
rich and the rest of society.

Some of the above findings may have been known on an anecdotal level, or
may seem obvious; our axioms allow us to quantify the forces at play and compare
different core-periphery partitions. For example, it is shown in [3] that also in the
well-accepted preferential attachment model, that founder cores might not satisfy
all our axioms. Thus, it is of a major interest to find evolutionary models in which
elites as described here emerge naturally.

Our results not only advance the theoretical understanding of the elite of social
structures, but may also help to improve infrastructure and algorithms targeted at
online social networks, e.g., to organize institutions better, or identify sources of
power in social networks in general.

Acknowledgements Supported in part by the Israel Science Foundation (grant 1549/13).

References

1. Alberich, R., Miro-Julia, J., Rosselló, F.: Marvel universe looks almost like a real social
   network. arXiv preprint cond-mat/0202174 (2002)
2. Andersen, R., Borgs, C., Chayes, J.T., Feige, U., Flaxman, A.D., Kalai, A., Mirrokni,
   V.S., Tennenholtz, M.: Trust-based recommendation systems: an axiomatic approach. In:
   Proceedings of the WWW, pp. 199–208 (2008)
3. Avin, C., Lotker, Z., Nahum, Y., Peleg, D.: Core size and densification in preferential
   attachment networks. In: Proceedings of the 42nd Int. Colloq. on Automata, Languages, and
   Programming (ICALP), pp. 492–503 (2015)
4. Avin, C., Lotker, Z., Peleg, D., Pignolet, Y.A., Turkel, I.: Elites in social networks: an axiomatic
   approach. http://bit.ly/2fqLPUT (2016)
5. Borgatti, S., Everett, M.: Models of core/periphery structures. Soc. Netw. **21**(4), 375–395
   (2000)
6. Borgatti, S., Everett, M., Freeman, L.: Ucinet: Software for Social Network Analysis. Analytic
   Technologies, Harvard (2002)
7. Borgatti, S., Everett, M., Johnson, J.: Analyzing Social Networks. Sage, London (2013)
8. Chung, F.R.K., Lu, L.: Complex Graphs and Networks. American Mathematical Society,
   Providence (2006)
9. Csermely, P., London, A., Wu, L.-Y., Uzzi, B.: Structure and dynamics of core/periphery
   networks. J. Complex Netw. **1**(2), 93–123 (2013)
10. Cucuringu, M., Rombach, P., Lee, S.H., Porter, M.A.: Detection of core-periphery structure in
    networks using spectral methods and geodesic paths. Eur. J. Appl. Math. **27**, 846–887 (2016)

11. Da Silva, M.R., Ma, H., Zeng, A.-P.: Centrality, network capacity, and modularity as parameters to analyze the core-periphery structure in metabolic networks. Proc. IEEE **96**(8), 1411–1420 (2008)
12. Dunbar, R.: Neocortex size as a constraint on group size in primates. J. Hum. Evol. **22**(6), 469–493 (1992)
13. Facundo, A., Atkinson, A.B., Piketty, T., Saez, E.: The World Top Incomes Database (2013)
14. Faust, K., Wasserman, S.: Blockmodels: interpretation and evaluation. Soc. Netw. **14**, 5–61 (1992)
15. Galeotti, A., Goyal, S.: The law of the few. Am. Econ. Rev. **100**, 1468–1492 (2010)
16. Geiger, D., Paz, A., Pearl, J.: Axioms and algorithms for inferences involving probabilistic independence. Inf. Comput. **91**(1), 128–141 (1991)
17. Hojman, D.A., Szeidl, A.: Core and periphery in networks. J. Econ. Theory **139**(1), 295–309 (2008)
18. Holme, P.: Core-periphery organization of complex networks. Phys. Rev. E **72**(4), 046111 (2005)
19. Mislove, A., Marcon, M., Gummadi, K.P., Druschel, P., Bhattacharjee, B.: Measurement and analysis of online social networks. In: Proceedings of the 7th ACM SIGCOMM Conf. on Internet Measurement, pp. 29–42. ACM, New York (2007)
20. Oxfam International: Working for the Few: Political Capture and Economic Inequality (2014)
21. Pareto, V.: The Mind and Society. American Mathematical Society, New York (1935)
22. Piketty, T.: Capital in the Twenty-First Century. Harvard University Press, Cambridge (2014)
23. Rombach, M.P., Porter, M.A., Fowler, J.H., Mucha, P.J.: Core-periphery structure in networks. SIAM J. Appl. Math. **74**(1), 167–190 (2014)
24. Rossa, F.D., Dercole, F., Piccardi, C.: Profiling core-periphery network structure by random walkers. Sci. Rep. **3**, 1467 (2013)
25. Yang, J., Leskovec, J.: Overlapping communities explain core–periphery organization of networks. Proc. IEEE **102**(12), 1892–1902 (2014)
26. Zhang, X., Martin, T., Newman, M. E.J.: Identification of core-periphery structure in networks. Phys. Rev. E **91**, 032803 (2015)
27. Zhou, S., Mondragón, R.: The rich-club phenomenon in the internet topology. IEEE Commun. Lett. **8**, 180–182 (2004)

Ranking Scientific Papers on the Basis of Their Citations Growing Trend

Michaël Waumans and Hugues Bersini

Abstract Analyzing databases of academic papers citations with the tools of graph and network sciences produced many different results in the past: Publications ranking algorithms, predicting the becoming of their popularity either using the citations only or in association with the co-author or affiliations networks, understanding better the "ethnological aspects" of citation practices. The examination of the dynamical properties of such networks, i.e. how their nodes in-degree grows in time, started more recently. In this paper, we propose a novel ranking algorithm that makes a key use of these growth characteristics (for instance rewarding young, emerging stars more, and old, declining ones less) while requiring much less information and computation. To validate our ranking results and compare them with more established algorithms such as PageRank and FutureRank, four well-known datasets are used.

Keywords Network analysis • Dynamic networks • Citations and bibliographic networks • Ranking • Prediction

1 Introduction

Citations networks have gained a considerable attention over the years. This growing interest may be justified by the crucial importance taken by citations in the evaluation of a researcher's career or of his academic professional progression. We are also attending a shift in attention where "publish or perish" is being gradually substituted with "be cited or perish".

Many different methods have been proposed to assess a researcher's influence such as the well known H-Index [1] and many variations [2, 3] on it, including the G-index [4], C-index [5] and E-index [6]. Different researches were also conducted on the evaluation of the current impact of an article as well as the evolution in time of its influence. This gives rise to the qualification of some

M. Waumans (✉) • H. Bersini
École polytechnique de Bruxelles, ULB, CoDE-IRIDIA, 50, Av. F. Roosevelt, CP 194/6, B-1050 Brussels, Belgium
e-mail: mwaumans@ulb.ac.be; bersini@ulb.ac.be

© Springer International Publishing AG 2017
E. Shmueli et al. (eds.), *3rd International Winter School and Conference on Network Science*, Springer Proceedings in Complexity, DOI 10.1007/978-3-319-55471-6_8

publications with exotic names such Rising Stars [7], Sleeping Beauties [8], Gems [9] or First-Movers [10]. Among the different algorithms that exist to estimate the importance of a publication, one finds PageRank [11]. Several modifications of this initial proposal [12–14] using simulations of network traffic have been proposed. Alternative approaches [15, 16] base their ranking on an attempt to predict what will be the most popular articles of tomorrow. The Z-Score [17] for instance presents good aptitude at an early identification of promising articles by measuring to what extent its citation growth differs from the others published in the same time frame. Some of these proposals use complementary networks like the corresponding authorship or affiliations ones to better anticipate the future fame of specific papers. Any paper will gain some ranks if published by eminent scientists, in well-known laboratories or simply referenced by popular authors. Other research concentrate on the theoretical modelling of such networks in order to better understand (an ethnological approach) the citation practice [18–20] and the observed temporal effects in bibliographic networks [21, 22]. An increasing number of algorithmic proposals examine these networks in a more dynamic fashion [23] while most of the existing approaches were limited to static snapshots: taking a picture of a network's topology at one point in time only. Among them, the ranking method presented in this paper is based on the type of citation growth of each article. By examining their citation growth dynamics, very recent articles that do present a high citation rate will be gratified a high ranking while old and well cited articles but with a decreasing citation rate will be disqualified.

Ranking algorithms such as PageRank, or those in the footsteps of it, rely only on an frozen-in-time topology of the network. In contrast, we only make use of the growth dynamics of the network by examining each article's time series. Not surprisingly, a dynamical trend is more adapted to predict the future than just a static snapshot. Using these growing trends, our proposal improves on the precedent static methods mainly to anticipate the fate of an article, with no need (such as FutureRank [24]) to use the co-author network or any additional information that present their own biases and drawbacks.

2   Dynamics

Four different datasets were used in this work: ArXiV TH, PH, APS [25] and PubMed.[1] Although the presented work was tested on all these datasets, we hereby focus on the Arxiv TH network, on which many ranking algorithms were tested in the past. Following the recent developments in citation network analysis,

---

[1] The ArXiV HEPTH and HEPPH datasets are available online at the following address http://www. cs.cornell.edu/projects/kddcup/datasets.html. The American Physical Society dataset is available online at the following address http://journals.aps.org/datasets. The PubMed dataset is available online at the following address http://www.ncbi.nlm.nih.gov/pmc/tools/ftp/.

Fig. 1 Examples of the most representative in-degree growth trends observed

we carefully examine the way the nodes in-degree grow in time. All scientific publications do follow different growths, but they also present similar trends. In a very first approximation, these trends can be loosely described as logarithmic, linear, exponential or sigmoid curves, while not exactly fitting any of these functions (Fig. 1).

2.1 Growth Trends

Since none of the observed growth curves perfectly fit any of the aforementioned functions, let's rather designate them using more neutral terms that resemble those of the Robert Penner's Tweening functions [26]. The five types do correspond to the following natural intuitions:

- **Growth-Out:** Papers presenting similarities to a logarithmic growth are losing the interest of the community. They still gather citations but less and less as time passes.
- **Growth-InOut:** This type of article starts by gathering few citations during the first months following its release, but gathers a lot more attention after a certain period of time. Following this considerable gain of attention, it starts losing this initial interest.
- **Growth-OutIn:** Some papers also follow a Growth-OutIn trend; they are however just a handful so we won't consider them in this work. They are articles that do gather a lot of citations but only later on, following a period during which they lost attention. They may be assimilable to sleeping beauties [8], but very few will keep this increase in interest at a constant pace for long.
- **Growth-In:** These papers gather a lot of interest and are cited more and more as time passes (similar to an exponential growth). Although this kind of article can be considered as stars in the field, ultimately they will also lose citations over the years.
- **Growth-Linear:** The articles characterized by a linear curve are on average presenting a constant growth. As a matter of fact, almost all articles are comparable to these Growth-Linear ones during the first months following their publication but again very few keep this constant pace for long.

The way scientific papers do gather citations over time changes and the identification of these dynamical characteristics and shifts poses various challenges. Specific features should indicate not just whether an article gains or loses citations but how. A small decline of the citation rate does not specifically mean that an article is losing its popularity, but a consequent one could. This detection consequently requires to better identify the growing dynamic regime in which an article falls at each point in time. In what follows, we propose a simple set of features allowing to better characterize the way each article in-degree citation rate changes over time.

2.2 Growth Characterization

These features are called quadrants although they differ from the analytical geometry definition and just refer to quarters of a well-defined area. A quadrant value defines the "amount of samples from a given normalized time series present in either one of the four quadrants defined by cutting the normalized space (i.e. X-axis and Y-axis) in two halves using the bisector and its perpendicular" (Fig. 2). This notion allows to clearly identify the shape of a given curve even better than a direct fitting would. More specifically, when following the evolution of a curve over time, the key transition from Growth-In to Growth-Out becomes more salient using quadrants as shown in the following examples. Different approaches were studied first to develop those features but none was as simple and efficient to detect the transitions from one dynamical regime to another.

Just using these four quadrants, Growth-Linear curves remain difficult to identify since they do oscillate among all of the four areas. In order to better identify this dynamical regime, an extra exclusion zone is added out of the quadrants. A point close enough to the diagonal going from $(0, 0)$ to $(1, 1)$ will be considered to be in this fifth "quadrant". By resorting to these five features, all the types of curve previously observed are clearly identifiable. Besides the capacity of this method to properly identify each type of growth, it can further characterize the way these types change in time.

Fig. 2 Definition of the four quadrants and the exclusion area around the diagonal

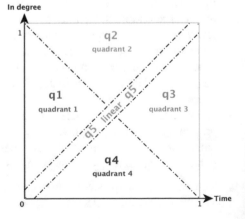

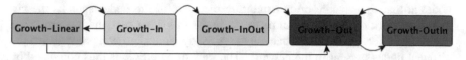

Fig. 3 Articles possible lifecycles

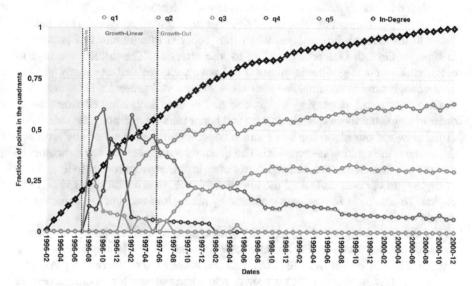

Fig. 4 Article AU6vljQ505SKIEIHnpJC evolution

The Fig. 3 also shows all possible transition patterns between the possible growth trends identified previously. The vertices of citation networks do not change over time. Once an article is published it may only be updated but its main content will not change and its popularity either: thus, the quadrant values computation are always done using an opening time window, considering the in-degree values for each month from the time of publication to the time chosen to make the computation.

We have so far proposed stable features to characterize the growth of any given curve. They allow to identify properly Growth-Linear, Growth-In, Growth-Out as well as Growth-InOut/OutIn trends and the transitions among those types of growth over time. The Fig. 4 illustrates this aspect. Looking at the evolution of the quadrant values over time, we can easily see that this specific article is first Growth-In to then become Growth-Out.

3 Ranking

Numerous ranking algorithms have been proposed over the years (e.g. Future Rank [24], CiteRank [13], various versions of PageRank [11], ...). The evolution in time is usually not taken into account in those proposals. These methods mainly use snapshots of an entire network at a given date and establish a ranking of the nodes at that time, while often exploiting metadata of the nodes such as the co-author network or the affiliation network. In the following sections, we propose a simple and effective algorithm that only uses the dynamical growth properties of the different nodes present in the network. The rank of an article is thus computed by using its in-degree time series only and not relying on the exact topology resulting from the connections made by the citations among the articles.

The most prominent articles usually have a high in-degree value. They belong either to the Growth-Linear type, when they succeed to maintain this popularity in time, or Growth-Out, when starting to lose attention. The articles we tend to better rank in our algorithmic proposal are young articles that obviously did not have enough time to accumulate many citations, but still present a Growth-Linear or Growth-In trend, displaying a high, constant or even increasing citation rate. In order to single out these specific articles and better rank them among the older stars of the network, our algorithm takes into account the growth trend of any article at the moment the ranking is computed. The quadrant values are the key indicators to achieve just that. In essence, the popularity of an article is proportional to its average in-degree growth over its last 12 months of existence, thus capturing its short term impact. This value is furthermore weighted by taking into account this same article growth trend.

As mentioned before, the quadrant values are used to characterize the growth trend but not its curvature. For instance, one Growth-Out article may have shown a high citation rate earlier in its life but a null one at the time of the ranking. Another one may still present a significant growth rate when the ranking is established. In brief, the first one would have left the stage while the second would still gather citations. It is then important to quantify the curvature of either the Growth-In or the Growth-Out curve to refine even more an article's rank. Our algorithm takes into account this *area* value. It is defined as the surface between the diagonals from $(0, 0)$ to $(1, 1)$ and the actual in degree time series. It may be positive for Growth-Out articles or negative in the case of Growth-In ones. (See Fig. 5 for an example of two Growth-Out curves. $area1 > area2$ that implies that the first article have lost more popularity than the second one.)

The necessary information for each article is computed from the in-degree time series as indicated in the algorithm at Lines 1-4. The indices indicate the time while n is the length of the time-series itself. The time period may be chosen as a year, a month or a day. Our analysis considers a 1 month time period to match the relatively small number of citations present in the datasets being used, thus restricting the time step we may use to keep consistent results. The final score of an article is computed by taking into account the rate at which it gathers citations as well as the trend of its growth curve. The final equation is presented below from Line 5 to 7.

$$inDegree = (inDegree_1, \ldots, inDegree_n) \tag{1}$$

$$growth_i = \sum_{j=0}^{i} inDegree_j \tag{2}$$

Fig. 5 Examples of Growth-Out curves with different area values

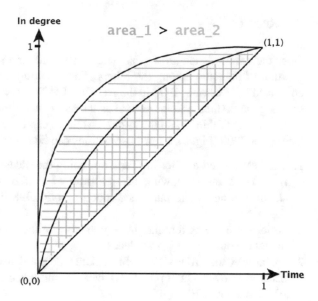

$$avgLast_i = \frac{\sum_{j=-12}^{-1} inDegree_j}{j} \qquad (3)$$

$$q1_i, q2_i, q3_i, q4_i, q5_i, area_i = compute(growth_i) \qquad (4)$$

$$scoreImpact_i = avgLast_i * (1 - \alpha) \qquad (5)$$

$$scoreTrend_i = avgLast_i * \alpha * (1 - area_i) * (q3_i + q4_i + q5_i) \qquad (6)$$

$$scoreTotal_i = scoreImpact_i + scoreTrend_i \qquad (7)$$

In the score computation, only three quadrant values are used: $q3$, $q4$ and $q5$. Our algorithm focuses on young articles displaying an increasing citation rate or older ones displaying a great and constant pace, so Growth-In and Growth-Linear articles, characterized by high $q3$ and $q4$ or $q5$ values. This emphasis will have for effect to increase the rank of the articles presenting such trends of growth. The sum of those quadrants is then multiplied by the average number of citations received over the last 12 months period prior to the ranking time, in order to capture the most up-to-date popularity of the publications. This effectively lowers the rank of older articles gathering less citations as time passes. As explained, the curvature of the time-series is estimated to further increase this aspect using the *area* value.

Our algorithm uses one parameter, α, that varies from 0 to 1. Maximising α increases the importance of the time series trend in the score (i.e. Algo. Line 6), while minimising it makes the score more focused on the average citation rate (i.e. Algo. Line 5). By adjusting this parameter, we can modify the weight given to the desired criteria. The results presented in the following are obtained with $\alpha = 0.5$.

4   Results

Once the score of each article is obtained, the rank of an article corresponds to the position of its value among the others. The resulting ranking from our algorithm computed in December 1998 in the ArXiV HEP-TH network is presented in Table 1 showing the first ten articles as well as the associated time series in Fig. 6. Looking at the ranking (Table 1) and the in-degree curves captured until December 1998 then December 2000 (Fig. 6), a few observations can be made :

1. The best ranked articles are far from being the oldest. The network being built from 1992, and following the intuition of Preferential Attachment [20], we should see an article published much before 1998 in the lead, but this is not the case.
2. Articles presenting a higher in-degree value at the time the ranking is computed do not occupy the very first places.
3. An article like AU6vlYJ805SKIEIHnms8, published in 1994, that starts to become Growth-Out in 1998, is clearly outranked by a more recent article: AU6vlivY05SKIEIHnpBW.
4. The article AU6vlIWw05SKIEIHnj8h does not present a high in-degree value by the year 2000. However, it displayed at the time of the ranking a surprising growth of interest with its trend becoming strongly Growth-In and its in-degree on average quadrupled. Such an article is definitely worth a closer look. However, its ranking went down during the year 1999 where this sudden increase in interest disappeared as quickly as it came. Anyway, this sudden spike in popularity explains its position in our ranking.

Those observations underscore the way our method does achieve its goal. Older articles that have lost attractiveness and have become Growth-Out do not occupy the best positions. Few of them that present an interesting trend, the Growth-OutIn ones for example, are not discarded and persist in being well ranked. Also, as it

Table 1 Ranking established in December 1998

Rank	ID	Publication date	1998–12	2000–12
1	AU6vlXcO05SKIEIHnmjQ	1997–11	664	1680
2	AU6vlivY05SKIEIHnpBW	1998–02	507	1234
3	AU6vlPZG05SKIEIHnk93	1998–02	468	1146
4	AU6vlezT05SKIEIHnoI0	1996–10	655	961
5	AU6vlQuR05SKIEIHnlNw	1995–10	491	691
6	AU6vlYJ805SKIEIHnms8	1994–07	864	1108
7	AU6vlaD305SKIEIHnnGR	1994–08	693	868
8	AU6vlIWw05SKIEIHnj8h	1992–05	208	296
9	AU6vlpGF05SKIEIHnqep	1995–10	657	974
10	AU6vlMlf05SKIEIHnkfZ	1995–03	806	984

The current in-degree value is indicated as a reference as well as the value in December 2000

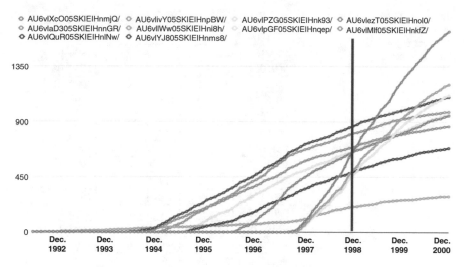

◇ AU6vlXcO05SKIEIHnmjQ/ ◇ AU6vlivY05SKIEIHnpBW/ ◇ AU6vlPZG05SKIEIHnk93/ ◇ AU6vlezT05SKIEIHnol0/
◇ AU6vlaD305SKIEIHnnGR/ ◇ AU6vllWw05SKIEIHni8h/ ◇ AU6vlpGF05SKIEIHnqep/ ◇ AU6vlMlf05SKIEIHnkfZ/
◇ AU6vlQuR05SKIEIHnlNw/ ◇ AU6vlYJ805SKIEIHnms8/

Fig. 6 In-degree curves at the time of the ranking both in Dec. 1998 and in Dec. 2000

was already proven in other research, the best article in terms of in-degree is not the best potential star of tomorrow anymore when not taking into account its growth trend. One question must nevertheless be answered, even though we aim at a slightly different objective: How does our algorithm compare to other ranking algorithms?

To validate the results of our proposal, we use the ranking produced by the algorithms FutureRank [24] (i.e. FR) and PageRank [11] (i.e. PR) as a basis of comparison. The PageRank algorithm is the traditional model with $\alpha = 0.9$ and random jumps taking place with probability of 0.1. The Future PageRank (i.e. FPR) is the PageRank score of articles based only on the citations starting in 2001 until the end of the dataset in 2003. At last, FutureRank is also based on PageRank. However, it aims at improving this ranking by predicting the future number of citations an article would receive. To achieve this goal, Sayyadi and Getoor have taken into account more information, mainly the publication time and the authors (by using a co-author network built using a strong matching criteria—the author's names must perfectly match). The top 20 articles according to FutureRank are presented in Table 2 with their corresponding Pagerank as well as the resulting ranking from our proposal.

The results presented in Table 2, not being directly compared to FutureRank by means of a rank correlation measure because of the different ranking criteria used, need to be discussed further to justify the validity of our approach. Few examples are detailed in the following, explaining the major differences between the results of our algorithm and the one proposed by Sayyadi and Getoor:

- **9711165:** This article, published in November 1997, is ranked 17 in FutureRank and 33 in Future PageRank. Our methods ranks this article in seventh position. In December 2000, this article presents the following features : $q3 = 0.30$, $q4 = 0.59$, $avgLast = 13.58$, $avgLife = 7.61$ and $area = 0.17$. The growth trend of

Table 2 TOP 20 articles ranked by FutureRank with their corresponding PageRank scores

ID	FR	FPR	PR	Proposal	Publication date
9711200	1	1	10	1	1997–11
9802150	2	4	28	5	1998–02
9906064	3	2	92	3	1999–06
9802109	4	5	37	6	1998–02
9908142	5	3	131	2	1999–08
9407087	6	10	1	29	1994–07
9610043	7	8	14	22	1996–08
9510017	8	14	4	20	1995–10
9711162	9	7	106	4	1997–11
9905111	10	6	174	10	1999–05
9503124	11	46	2	79	1995–03
9408099	12	25	6	76	1994–08
9510135	13	77	13	54	1995–10
9510209	14	9	40	8	1995–10
9611050	15	29	76	78	1996–11
9409089	16	15	95	42	1994–09
9711165	17	33	211	7	1997–11
9204099	18	62	69	107	1992–05
9410167	19	94	5	210	1994–10
9603142	20	17	167	11	1996–03

Ranking established at the very end of the year 2000 (*PR* PageRank, *FR* FutureRank, *FPR* Future PageRank)

the article is Growth-In and its popularity started to increase more recently. It is not a very recent article and it did not gather much attention before 1997, but recently, it has attracted more and more attention.

- **9407087:** This article, published in July 1994, is ranked 6 in FutureRank and 10 in Future PageRank. Our method ranks this article in 29th position. At the time the ranking is established, its features are : $q2 = 0.45$, $q3 = 0.0$, $q4 = 0.22$, $avgLast = 9.25$, $avgLife = 13.69$ and $area = 0.05$. This paper is of the Growth-Out trend since the end of the year 1999 and still loses citations as time passes. However, this trend is not well marked, and its popularity slowly decreases. This article is just old and simply looses the focus of attention.

- **0005031:** A very young paper published in 2000–05 is ranked in 15th position in our proposal while not reaching the top 20 in the other ones (e.g. The most recent article in the top 20 is from 1999–08 in the proposal from Sayyadi et al.). PR places it at the 1993th position. It has a $avgLast = 9.77$ and displays a strong Growth-In trend. This article, although very recent, has a very strong impact since its publication. To put this in relief, the 20th best article overall in term of $avgLast$ receives only 8.25 citations whatever their age.

- **9204099:** This article is another interesting one in our ranking. Even though it was published in May 1992 and could be ranked lower since our algorithm emphasizes the most recent ones, it is still ranked at the 107th position of more than 20,000 papers. This publication was gathering citations slowly since its publication, at a rate of two citations per month on average. However, it displayed a sudden regain of interest by the start of year 1998, entering so the Growth-OutIn trend: an old article regaining a lot of attention long after its publication. It is only ranked in 107th position because this sudden interest was lost by 1999.

The FPR results can be seen as serving a sort of "ground truth" for the network's future traffic. Therefore, it is relevant to directly compare our proposal to it. Looking at Table 2, we see that our results do resemble more closely the FPR results. The Spearman's rank correlation coefficient between the future PageRank computed using the citations produced between 2001 and 2003 and our proposal calculated in 2000 is 0.65.

5 Computational Aspects

From a purely computational perspective, our algorithm presents some advantages.

First, it is a non iterative method that only depends on one article's time series, not on the whole network's topology. This approach, while reaching its goal of improving young paper's ranks, only relies on the time series extracted from a citation network. Thus, the amount of data to gather, disambiguate and validate, is reduced compared to other proposals making additional use of affiliation or co-author networks. However, using alternative sources of information proved to be helpful in improving the predictive capabilities of other algorithms and could be equally used to refine our proposal. That said, by focusing on the citation dynamics only, we have proposed a ranking that gives a better chance to young publications while ignoring the identity of the author or laboratory issuing the publication.

Second, since the rank of an article is only dependent on the characteristics of its growth, adding new vertices or edges to the network only requires to run the computations again on the modified vertices, and not the entire network's topology. It may be even further improved by avoiding unnecessary computations for articles receiving very few citations. Other papers show different possible optimizations [27, 28] of the PageRank algorithm and its followers by approximating the traffic in the simulations or by focusing on local computations. Our algorithm hereby only needs to update the required article rank while not making approximations and still keeping the method simple and easily scalable to handle larger networks.

Our proposal is then perfectly suited for the computation of rankings on fractions of the network to be used in more complex scenarios (e.g. Application in [29] to build Genealogical Trees live from the Scopus API). It is useable as well on very large datasets for which the amount of metadata may be limited to a minimum using the citations and publications dates only. We thus avoid the trouble of dealing with multiple networks of multi-million nodes and the challenge of properly weighting the edges used in traffic simulations.

6   Discussion and Conclusion

The ranking method proposed here dissociates the influence of the age of an article from its actual impact. We consequentially provide a new ranking algorithm that allows to emphasize the short term impact. This algorithm, by relying on the in degree time series only, alleviates the burden of all the metadata validation coming from multiple sources of information. As explained, our algorithm uses an analysis of the in-degree growth curve through temporal windowing until the time the ranking is computed. This was chosen because of the particular dynamics of citations networks. Our proposal could however be applied on other types of more dynamic network such as social ones by using a sliding time-window. This would however imply the usage of other features than the quadrants. These specific features are perfectly adequate in this use case but probably not in others since the existing growth trends in such networks could be different.

Our algorithm also offers good predictive capabilities since it takes into consideration the time of publication as well as the evolution of the rate at which any article keeps receiving citations. This approach, though simple, only uses the citations time series obtained by each article over time. By looking closely at the dynamics of individual articles rather than taking into account the fame of the writers or laboratories involved, our algorithm succeeds to reach its goal. It keeps ignoring other sources of information and thus leveraging the induced complications of building multiple networks that may sometimes be rather heavy to handle.

References

1. Hirsch, J.E.: An index to quantify an individual's scientific research output. Proc. Natl. Acad. Sci. **102**(46), 16669–16572 (2005). doi:10.1073/pnas.0507655102
2. Batista, P.D., Campiteli, M.G., Kinouchi, O.: Is it possible to compare researchers with different scientific interests? Scientometrics **68**(1), 179–189 (2006)
3. Sidiropoulos, A., Katsaros, D., Manolopoulos, Y.: Generalized Hirsch h-index for disclosing latent facts in citation networks. Scientometrics **72**(2), 253–280 (2007)
4. Egghe, L.: Theory and practise of the g-index. Scientometrics **69**, 131 (2013)
5. Bras-Amoras, M., Domingo-Ferrer, J., Torra, V.: A bibliometric index based on the collaboration distance between cited and citing authors. J. Informet. **5**(2), 248–264 (2011)
6. Zhang, C.-T.: The e-index, complementing the h-index for excess citations. PLoS ONE **4**(5), e5429 (2009). doi:10.1371/journal.pone.0005429
7. Li, X.-L., Foo, C.S., Tew, K.L., Ng, S.-K.: Searching for rising stars in bibliography networks. Database Syst. Adv. Appl. **5463**, 288–292 (2009)
8. Ke, Q., Ferrara, E., Radicchi, F., Flammini, A.: Defining and identifying sleeping beauties in science. Proc. Natl. Acad. Sci. U. S. A. **16**, 7426–7431 (2015)
9. Chen, P., Xie, H., Maslov, S., Redner, S.: Finding scientific gems with Google. J. Informet. **1**, 8–15 (2006). doi:10.1016/j.joi.2006.06.001
10. Newman, M.E.J.: The first-mover advantage in scientific publication. Europhys. Lett. **86**(6), 68001 (2009)
11. Brin, S., Page, L.: The anatomy of a large-scale hypertextual web search engine. Comput. Netw. ISDN Syst. **30**(1–7), 107–117 (1998)

12. Krapivin, M., Marchese, M.: Focused PageRank in scientific papers ranking. In: Digital Libraries: Universal and Ubiquitous Access to Information, pp. 144–53. Springer, Berlin/Heidelberg (2008)
13. Walker, D., Xie, H., Yan, K.-K., Maslov, S.: Ranking scientific publications using a simple model of network traffic. J. Stat. Mech. **0706**, P06010 (2007)
14. Radicchi, F., Fortunato, S., Markines, B., Vespignani, A.: Diffusion of scientific credits and the ranking of scientists. Phys. Rev. E **80**, 056103 (2009). doi:10.1103/PhysRevE.80.056103
15. Ghosh, R., Kuo, T.-T., Hsu, C.-N., Lin, S.-D., Lerman, K.: Time-aware ranking in dynamic citation networks. In: Proceedings of the 2011 IEEE 11th International Conference on Data Mining Workshops, pp. 373–380 (2011). doi:10.1109/ICDMW.2011.183
16. Yao, L., Wei, T., Zeng, A., Fan, Y., Di, Z.: Ranking scientific publications: the effect of nonlinearity. Sci. Rep. **4**, Art:6683 (2014). doi:10.1038/srep06663
17. Newman, M.E.J.: Prediction of highly cited papers. Europhys. Lett. **105**, 28002 (2014)
18. de Solla Price, D.J.: Networks of scientific papers. Science **149**, 510–515 (1965)
19. de Solla Price, D.J.: A general theory of bibliometric and other cumulative advantage processes. J. Am. Soc. Inf. Sci. **27**, 292–306 (1976)
20. Barabasi, A.-L., Albert, R.: Emergence of scaling in random networks. Science **286**, 509–512 (1999). doi:10.1126/science.286.5439.509
21. Medo, M., Cimini, G., Gualdi, S.: Temporal effects in the growth of networks. Phys. Rev. Lett. **107**, 238701 (2011). doi:10.1103/PhysRevLett.107.238701
22. Eom, Y.-H., Fortunato, S.: Characterizing and modeling citation dynamics. PLoS One **6**, e24926 (2011)
23. Wang, D., Song, C., Barabasi, A.-L.: Quantifying long-term scientific impact. Science **342**, 127–132 (2013)
24. Sayyadi, H., Getoor, L.: FutureRank: ranking scientific articles by predicting their future PageRank. In: SIAM International Conference on Data Mining SDM09 (2009)
25. Redner, S.: Citation statistics from 110 years of physical review. Phys. Today **58**, 49 (2005)
26. Penner, R.: Programming Macromedia Flash MX. Dynamic Visuals, Part 3. Osborne, Berkeley (2002). ISBN13:978-0072223569, ISBN10:0072223561, http://robertpenner.com/easing/
27. Bressan, M., Peserico, E., Pretto, L.: Approximating PageRank locally with sublinear query complexity. ArXiV (2014)
28. Borgs, C., Brautbar, M., Chayes, J., Teng, S.-H.: Multi-scale matrix sampling and sublinear-time PageRank computation. Internet Math. (2013). doi:10.1007/978-3-642-30541-2_4
29. Waumans, M., Bersini, H.: Genealogical trees of scientific papers. Plos One (2016). doi:10.1371/journal.pone.0150588

Towards Network Economics: The Problem of the Network Modus of Value

Alexey A. Baryshev

Abstract Network paradigm is already on the threshold of economic science. This paper addresses conditions for the reality of network economy and the genesis of the network content of value as a base category of economics. Identification of the specifics of network phenomena as a perspective cornerstone of society and economy is carried out. The concept of ontological status is employed as methodological optics for the determination of the maturity of the network content of social and economic categories. The network modus of value is hypothesized on the basis of "affordance" and "preferential attachment".

Keywords Network economy • Austrian economic theory • Economic action • Heterogeneity of value • Shared value • Affordance • Preferential attachment

1 Introduction

Development of information technology actualizes ideas and understanding of network economy as quite a new mode of production of conditions for human life. Simultaneously, theoretical comprehension of ideas, which are about to occur and which therewith have been turned into an element of everyday knowledge, is a challenging task.

The rapid rise of prosperous corporations Google (Alphabet), Facebook, Yahoo and other giants of the IT industry and their approaching the previously inconceivable capitalization of a trillion dollars create anxious expectations of radical global changes in the economic order [1]. These significant quantitative changes in the amount of the working capital are related not merely to fundamental transformations of methods of value creation [2]. The thing is that the base economic categories (value, interest rate, etc.) will continue to exist as long as economy exists regardless of the fact whether it is network economy or whatever else. However, this does not mean that their content will remain the same. These changes affect the fundamentals

A.A. Baryshev (✉)
Laboratory of Big Data and Problems of Society, National Research Tomsk State University, pr. Lenina 36, 634050, Tomsk, Russia
e-mail: barishevnp@mail.ru; baryshevnp@gmail.com

© Springer International Publishing AG 2017
E. Shmueli et al. (eds.), *3rd International Winter School and Conference on Network Science*, Springer Proceedings in Complexity, DOI 10.1007/978-3-319-55471-6_9

of the economic system, viz., the categories *value* and *surplus value*. With this in mind, we intend to show that network economy changes the content of basic economic categories.

The paper focuses on the ontological status of social networks in economic objects, correlated with the stages of formation of both the theory of social networks, and the formation of the network content of economic categories. By "ontological status", we mean the degree, to which network processes stipulate the key characteristics of the behavior of an object in general. In other words, the ontological status of one or another phenomenon characterizes the mechanism of its self-maintenance, i.e., whether the phenomenon can exist independently on the basis of those properties, which are defined as essential, or needs constant support from the outside.

Hence, for example, the existence of a natural reserve, unlike biocenosis in wild nature, depends on decisions of the state environmental agencies. Therefore, the ontology of nature processes in the framework of the reserve has a dependent or *weak* status.

In this case, networks creating biocenosis can coexist within a hierarchy provided by governmental control. If biocenosis is funded by means of revenues derived from tourist activities performed within it; then, the ontology of nature networks becomes dependent on a substantially different base, viz., market of tourist services.

According to the main assumption of our article, an object can be supported on the basis of various principles due to network interactions, hierarchy and market. Consecutively, the ontological status of an object can be determined based on any of these principles. As an illustration, an industrial organization of classical type acts as a hierarchical system in spite of the fact that plenty of informal interactions among people within the organization are carried out in network mode. If we consider this object from the viewpoint of hierarchy; then, it is characterized by *strong* ontological status. On the contrary, the ontology of social networks can be defined as *weak* or precarious. At the same time, the considered organization can be investigated as a network, but not as a hierarchy. However, it is necessary to remember that its existence is mainly fulfilled on the basis of hierarchy, and the network in this case is a research representation of the object. Regardless of the degree of clarity of this representation, it is not able to change the real ontological status of the network in the course of reproduction of this object.

In this regard, the following questions are considered in the paper:

1. How does the network view correspond to the network organization of economic processes?
2. What conditions are necessary for transformation of networks into the basis of economy?
3. What in this case happens to the content of basic economic categories, especially, to value?

2 Weak Ontology of Networks

The theory of social networks began to emerge in the late 19th —early 20th centuries due to the interest of sociologists in the formation of groups of people. In the 1930s, the first analytical tools for description of interpersonal relationship patterns in a small group appeared. They were developed by J. Moreno [3, 4]. The term "social network" was first suggested by James Barnes in 1954 and was defined as " ... not a corporate body, but rather a system of social relations through which many individuals carry on certain activities, which are only indirectly coordinated with one another" [5].

The subject of interactive processes in human communities, which were characterized by flexibility, adaptability, and absence of a single center and clear boundaries, was a research domain for anthropology, psychology and the sociology of small groups called interactionism.[1]

The relationist focus of these studies on individual interactions and relations among people surprisingly correlated with the analytical apparatus developed by mathematicians (Erdős and Rényi) to describe social networks that did not have obvious principles of construction. It was critical to regard networks as non-deterministic formations.

These approaches to understanding networks contrasted with the traditional understanding of society as a stable entity. As a result, the increased attention shifted from the attributive characteristics of individuals to their relationships. However, the principles of methodological individualism and interactionism proved ineffective to support social macro processes.

B. Latour demonstrated that in order to create a society based only on interactions, individuals would have to dedicate all their time to it [6]. Therefore, the status of networks as a substance that ensures the unity and stability of society, turned out to be precarious from the ontological point of view.[2]

The economic theory of that time also attempted to comprehend the structure and functioning of society and market on the basis of interactionism and individualism.

---

[1]Interactionism is a theoretical perspective, according to which all social phenomena can be comprehended on the base of human interactions. It arose as rejection of the structuralist approach to understanding social phenomena in terms of their relation to a larger structure or through addressing social phenomena of a higher level. Interactionism can be interpreted as early manifestation of network methodology in social sciences due to similarity of their major principle, that is to say, there is nothing but individuals (nodes) and relations (links) between them.

[2]In order to secure interactions and networking from structuralist assumptions of privileged structures that are responsible for stability and integrity of society and social objects, B. Latour substituted individuals as nodes for "actants" of heterogeneous networks, in which the distinction between "humans" and "non-humans" was eliminated. This is the core of the Actor-Network Theory (ANT) developed by Latour and his colleagues. Due to the ANT, interactionism ("intersubjectity" in B. Latour's terms) was transformed into "interobjectivity". The place of the ANT in the general evolution of the network approach to society will be considered in our next paper.

The Austrian school of economic theory designated the entrepreneur as the principal agent of formation of economic categories and regarded market as an information system, a sphere of distributed knowledge [7]. On the one hand, the activity of the entrepreneur was presented as an absolutely interactive and cognitive process. On the other hand, the value was interpreted by the Austrians in purely physical terms on the basis of utility.

The most important concepts characterizing activities of the entreprencur became entrepreneurial alertness [8], opportunity discovery [9], and reshuffling resources along with their newly discovered properties [10]. Unlike the first two concepts, reshuffling relates to network matter and can be considered a prototype of network approach [11]. However, due to the naturalistic background of the Austrian economic theory [12], the concept of the objective value (market price) reveals no traces of the mentioned network properties. Consequently, social value is interpreted as a physical characteristic of goods.[3] Therefore, transition from the world of utilities and human needs, which cannot be separated from the world of the physical existence of man, to values is the most vulnerable part of the Austrian theory.

Apparently, attempts of the examined economic and social schools to theoretically describe society and market on the basis of individualism and interactionism failed in terms of gnoseology because of the conflict between their network ideology, on the one hand, and, on the other hand, because of the idea that networks are something external in relation to things, which are "placed" inside networks. In terms of ontology, the attempts were not productive due to the fact that apart from the mechanism of interactions between individuals, society also employs some other means of maintaining stability and integrity.[4] This focuses us on analysis of the heterogeneous bases of social and economic life. One of these bases is networks. The bases are usually applied to qualitatively different historical periods of economy and society. As an example of this, D. Bell's axial principles of preindustrial, industrial and post-industrial societies are to be noted [13]. However, in fact, these principles are not necessarily designated with reference to time and space, creating independent institutional realms [14]. In re bases that create institutionally different organizational integrities, we share W. Scott's approach.

According to W. Scott's theory of bases and carriers of institutionalization, there are regulative, normative and cognitive "pillars" of institutions [15]. These pillars

---

[3]The Austrian School's physical interpretation of value implies that they deny a specific metric of the world of values. They understand value based on conventional metric and logic of interactions of isolated things, viz., interactions between the human and the good, specifying it as subjective utility. For instance, the concept of distance, identified as a ratio of route to a particular standard of length, appears to be inapplicable for the purpose of measuring distance in network reality; likewise, the concept *subjective utility* is inapplicable in economics, which is regarded as network.

[4]It is possible to say that contrary to the statements of the Austrians, the functioning of their "network" in an implicit form assumes something else except for individuals interacting with their subjective interests and estimates. In this regard, their metaphor "kaleidics" (from "kaleidoscope") [23, 24] is not a synonym, but a complete antithesis of network, as it implies availability of a special hidden mirror device that creates an instant change of the multi-colored picture.

are simultaneously represented in market economy, but the normative pillar plays the leading role. This means that description of the functioning of such economy in the subjectivist manner of the Austrian School is at least untimely. In fact, untimely ascription of the cognitive pillar to market categories brought the Austrian theory to a vision of economy as a self-organizing, dynamic and interactive information system, which contradictorily coexisted with their naturalistic understanding of the categories.

3 The Middle Network Ontology of Social and Economic Phenomena

After the 1970s and especially the 1980s, the general theory of networks accumulated a critical mass of new knowledge, formulated with the universal language of mathematics. The general network metrics theory [16, 17] was deepened, metrics of social objects [18, 19] were developed, "small world" concepts [20, 21] and concepts of structural holes [22] were furthered. Networks finally ceased to be represented as a kind of metaphysical essence and metamorphosed into a construction principle for the entire society and economy—not only local objects. This paradigm shift was stipulated by the discovery of a new class of network, i.e., free-scale networks [25].

The new concept enables us to reconsider network theory and to integrate numerous theories of social objects based on individuals' interactions [26–28]. The ontological base of this breakthrough is the development of new human communication methods and a factual increase in the importance of individual actors in socioeconomic processes. In economics, this was clearly manifested in a phenomenon that was first observed in the 1970s, viz., the so-called sustainable competitive advantages revealed in a range of successful companies. An explanation of sustainable competitive advantages emerged in the strategic theory of the firm [29] that identified them as presence of valuable, rare, inimitable and non-substitutable resources (VRIN-resources) [30].

Economists of the Austrian school made the greatest progress in explaining the above-mentioned phenomena by developing the concept of the entrepreneurial organization [31, 32], which lied in the framework of subjectivism and radical subjectivism[5] [33]. They also developed the concept of capital heterogeneity into the concept of human capital heterogeneity, or heterogeneity of mental models

---

[5]Radical-subjectivist strand of the Austrian Economics considers future unforeseeable. Compared to the common subjectivism of the Austrian school, radical subjectivism does regards market not only as a procedure of discovery (of future) but also as a process of dynamic creation (of future). There are three levels of subjectivism forming the radical-subjectivist stance "First, the subjectivism of wants ... Second, the subjectivism of ends and means ... Finally, the subjectivism of active minds recognizes that in all aspects of action the active mind may produce interpretations and possibilities the observing economist cannot imagine in advance [39]."

of entrepreneurial team participants [34]. An entrepreneur's main work task was already reshuffling (that seems to be analogous to "reassembling" of the Actor Network Theory) of people's cognitive properties that resulted in obtaining a unique combination of human capabilities, which could turn into the company's VRIN-resource and its invulnerability under conditions of competition. Despite this theoretical advance, the radical-subjectivist approach left unchanged the content of the basic economic categories although the idea of revising the concepts "good" and "value" was about to occur; for example, in ANT-based economics of qualities [35].

Attempts to develop a new vision of value include the theory of co-creation of value [36], the concept of shared value [37], and the concept of value constellation [38]. Problematization of value topology is another major achievement. We, however, should note that these endeavors addressed networks merely as methods of generating value in the conventional sense of the word. Therefore, in these cases, value itself is based (as before) on such a property of reality of the human physical world as utility, which is more personalized now and assumes greater involvement of the consumer in the process of value creation. In addition, it is more distributed among institutionally and technologically heterogeneous manufacturers, and more comprehensive because it embodies the result of the integrated company's activity. An integrated company has a value-creation system where not only suppliers and consumers, but also many other business partners and allies collaborate.

4 Conditions for the Reality of Network Economy and Hypotheses of Modi of Value

For the purpose of formulating a content of economic categories, adequate to network reality, we ought to clarify conditions for the reliable (or strong) ontological status of *any* reality. The following three statements are true with regard to any phenomenon being a steady and reliable object of a certain reality. Firstly, any object supports the existence, interacting with other objects of this reality. Secondly, the nature of the interactions is set by specifics of these particular objects. Thirdly, the measurement and observation tools of the studied objects are homogeneous to them and their interactions. The mechanical picture of the world, in which objects, interactions (forces) and means of cognition are consubstantial with respect to each other, is a clear illustration of this rule. It is evident that networks should correspond to the stated conditions.

With this in mind, it is possible to say that network economy arises when and where there is a particular *network trinity*, which consists in the following:

- Networks in the form of the world-wide web become carriers and transmitters of economic agents' actions, a domain, in which they interact, viz., their *network space*.
- Networks become flexible, alternative and mobile forms of human relations, forming a new form of a collectivity, i.e. *social networks*.

- Networks in the form of computer programs of machine learning, visualization of objects of a new type and other network facilities become adequate tools of cognition the new reality based on a universal *network science* [40].

Network trinity, being the perfect form of the cognitive base for human relations, modifies the representation of the social and, respectively of the content of social and economic categories. Indeed, in the case of the regulative pillars, the social is the authorized by the power matter or the sacral; in the case of normative pillars, the social is a reified spontaneous force of collective rationality. Then as now, the social is a network matter under conditions of specified *triunity of networks*.

Now, we can suggest a draft research program, which describes the main problems to be solved, and a number of hypotheses. The program comprises the following stages: definition of economic value in the network world; description of the process of market value transformation into network value and interaction between them; consideration of the network world's heterogeneity; study of the distinguishing features of the processes of value creation within specific parts of network reality and interactions between nodes, belonging to various parts of this fragmented reality; investigation of interactions between economic categories bearing a formed network content and economic categories, in which the content is created by market or coercive methods. Within this framework, the first most general hypothesis is to be formulated and then the subsequent hypotheses are to be derived from it.

H-1. Economic action is connected with the value of the product created by this action because its structure includes procedures for convincing people that this product makes sense for their life.

Support. Economic value is an economic duplicate of the social category of sense. As products are increasingly moving away from "natural needs", it is becoming more obvious that their infusion with human life requires creation of certain convictions in human minds. This implies convincing people that possession or consumption of a product makes sense. Different methods for convincing will result in various modes of economic value.

H-2. In the absence of special means of convincing, the value of products in the isolated accidental interaction "seller-buyer" is created through the rhetorical component of the economic action.

Support. The linguistic turn in economics revealed rhetoric as an essential part of economic life [41]. This part is connected with the semantic, valuable content of economic processes. If not a seller's capability of convincing a would-be buyer of his goods' special effect on the buyer's life, what else can determine the price when a seller and a buyer meet for never-before-seen goods? The modus of value arising in this case is rhetorical.

H-3. In the case of domination of hierarchical relations in an economic system, the modus of value is stipulated as coercive by the economic action for implementation of instructions on quality and quantity of goods that can ensure the matching of buyers' convictions related to goods relevance and their imputed ideas about the proper life.

Support. Based on anthropological and ethnographic investigations, Polanyi showed that market value is absent from traditional societies [42]. In today's economy, the value of goods is composed of actions oriented to compliance with regulations. This stance is an economic duplicate of the philosophical idea of sense, according to which the rational action is an action conforming to rules [43].

H-4. If methods for convincing consumers of product' value based on power and tradition are weak, if cognitive methods cannot work on a public scale; the modus of value is established as market modus through collective reified interaction among all producers and all potential buyers.

Support. The life of methods for convincing consumers of product value based on reified collective rationality is still an insignificant episode (having transitional nature) in human history. The transitivity of the market modus of value is determined by the fact that it creates a standard of human life, denying the value of an individual life trajectory. As soon as human identity commences to play a significant role in the creation and functioning of economy, the personalized modus of production that assumes direct interaction of producers and consumers [44] emerges.

H-5. Under the conditions of mass production of innovations, new products' capacity for further participation in humans' life is caused by matching the already existing products with the newly-invented ones. In other words, their affordance [45] serves as a means of convincing people of products' value, thereby, the affordance modus of value arises.

Support. Value as well as sense is created in context. Capacity for affordance (including technology affordance [46]), openness for linkage with a multitude of other things or technologies creates the value of particularly new products.

The market modus of value implicitly assumes the presence of "natural needs and wants", the order and the scope of which are a certain reality. Incidentally, mass innovation products satisfy more and more "artificial" needs; therefore, they (products) must convince people of their value by means of creating their own context, where they make sense. The order and scope of needs in that case become a matter of reassembling or reshuffling, but not the natural reality of the needs. Reassembling leads to emergence of product matching networks, concentrating around products that bear the highest value of affordance. The metrics of this network is now replacing the former "natural" ordinality of goods. The value of affordance is the precursor to network value. In addition, the network specifics of affordance consist in the fact that it is performed in the absence of *network trinity*. This makes it a dependent (not self-sufficient) modus of value as it must rely either on market modus of value or on the coercive modus of value.

H-6. Providing that the three conditions of reality of network economy (*network trinity*) are observed within product creation, the product value is created on the ground of preferential attachment [47], i.e. the preferential attachment modus of value operates here.

Support. Preferential attachment allows considering the value of network economy products in reliance on the degree of preferentiality of attachment. Both the utility and abstract labor doctrines include particular products into the scope of goods that ensures satisfaction of relatively static needs in the framework of the market modus of value; preferential attachment involves products in a dynamically developing network. In this respect, the network modus of value and the market modus of value perform the same function of product integration into the human life world. The network modus of value or the preferential attachment value is similar to affordance value. However, unlike the latter, the network modus of value must be capable of integrating all the antecedent modi; thereby providing their subordination to the processes that occur in the purely network-based economic space.

5 Conclusion

The conducted study proves that development of network theory is part of the process of acquisition of a reliable (or strong) ontological status by network phenomena. The weak ontology of networks is manifested in theories that allow obtaining a network-based technical tool for description of society. When social networks commence to declare themselves as real applicants for the role of dominant sociality carriers based on the cognitive pillar, the middle ontology is achieved. However, social and economic theories usually exaggerate the significance of network effects in society and economy.

A criterion for a reliable ontology of network economy is formation of the actual network content of habitual economic categories. The ad hoc factor is network transformation of the basic elements of economic reality (i.e. objects, interactions among them, and tools for man's cognitive and practical use of them), whose unity we named *network trinity*.

The value of network economy products under these conditions is determined neither on ground of utility nor abstract labor or relevance to imperious instructions or traditions. In our view, value formation occurs under conditions of pure network economy on the base of preferential attachment. Consequently, it is possible to retrospectively propose the heterogeneity of modi of value. The concepts of such modi of value as market value, coercive value, affordance value, preferential attachment value enable us to formulate a vision of network economy as of economy based upon the value of preferential attachment, on the one hand; but on the other hand, representing a patchy field of interactions of all the above-mentioned modi.

The most important problem of the theoretical cognition of the network modus of value and its interaction with the antecedent modi is the problem of money as an economic category. Judging by the antagonism that the official society shows in response to bitcoins as a form of network money, it is obvious that regulative and normative economic pillars do not intend to simply surrender to the cognitive pillar in the form of networks along with their new concept of value.

References

1. Zuboff, S.: The secrets of surveillance capitalism. In: Frankfurter Allgemeine Zeitung. http://www.faz.net/aktuell/feuilleton/debatten/the-digital-debate/shoshana-zuboff-secrets-of-surveillance-capitalism-14103616-9.html?printPagedArticle=tru (2010). Accessed 10 Oct 2016
2. Zuboff, S.: Creating value in the age of distributed capitalism. In: McKinsey Quarterly. http://www.mckinsey.com/business-functions/strategy-and-corporate-finance/our-insights/creating-value-in-the-age-of-distributed-capitalism. Accessed 10 Oct 2016
3. Moreno, J.L.: First Book on Group Psychotherapy, 3rd edn. Beacon House, Beacon (1957)
4. Moreno, J.L.: Who Shall Survive? Foundations of Sociometry, Group Psychotherapy and Sociodrama. Beacon House, Beacon, NY (1953)
5. Barnes, J.A.: Class and committees in a Norwegian Island Parish. Hum. Relat. 7, 48–49 (1954). http://pierremerckle.fr/wp-content/uploads/2012/03/Barnes.pdf. Accessed 23 Oct 2016
6. Latour, B.: Une sociologie sans objet? Note théorique sur l'interobjectivité. Sociologie du travail. **36**(4), 587–607 (1994)
7. Hayek, F.A.: Economics and knowledge. A presidential address to the London Economic Club, 10 Nov. 1936. In: Library of Economics and Liberty. http://www.econlib.org/library/NPDBooks/Thirlby/bcthLS3.html (1937). Accessed 23 Oct 2016
8. Foss, N.J., Klein, P.G.: Alertness, action, and the antecedents of entrepreneurship. J. Priv. Enterp. **25**(2), 145–164 (2010)
9. Klein, P.G.: Opportunity discovery, entrepreneurial action, and economic organization. Strateg. Entrep. J. **2**(3), 175–190 (2008)
10. Lachmann, L.: Capital and its Structure. Sheed, Andrews, & McMeel, Kansas City (1956)
11. Birner, J.: Introduction. In: Jack Birner, J., Garrouste, P (eds) Markets, Information and Communication. Austrian Perspectives on the Internet Economy, pp. 5–17. Routledge, London, (2004)
12. Birner J FA Hayek: the radical economist. New York University University, Department of Economics.. http://econ.as.nyu.edu/docs/IO/28047/Birner.pdf. Accessed 20 Oct 2016
13. Bell, D.: The Coming of Post-Industrial Society: A Venture of Social Forecasting. Basic Books, New York (1973)
14. Waters, M.: Daniel Bell (Key Sociologists). Routledge, London (1996)
15. Scott, W.R.: Institutions and Organizations: Ideas and Interests, 3rd edn. Sage Publications, Los Angeles (2008)
16. Newman, M.E.J.: The physics of networks. Phys. Today. **61**(11), 33–38 (2008)
17. Albert, R., Barabási, A.L.: Statistical mechanics of complex networks. Rev. Mod. Phys. **74**, 47–97 (2002)
18. Wasserman, S., Faust, K.: Social Network Analysis: Methods and Applications. Cambridge University Press, Cambridge (1994)
19. Wasserman, S., Robins, G.: An Introduction to random graphs, dependence graphs, and p*. In: Carrington, P., Scott, J., Wasserman, S. (eds) Models and Methods for Social Network Awnalysis, pp. 148–161. Cambridge University Press, Cambridge, New York (2005)
20. Watts, D.J., Strogatz, S.H.: Collective dynamics of "small-world" networks. Nature. **393**, 440–442 (1998)
21. Fouler, J.H.: Turnout in a small world. In: Zuckerman, A. (ed) The Social Logic of Politics: Personal Networks as Contexts for Political Behavior, pp. 269–287. Temple University Press, Philadelphia (2005)
22. Burt, R.: The social capital of structural holes. In: Guillen, M.F., Collins, R., England, P., Meyer, M. (eds) New Directions in Economic Sociology, pp. 201–246. Russel Sage Foundation, New York (2001)
23. Lachmann, L.M.: From Mises to Shackle: an essay on Austrian economics and the Kaleidic society. J. Econ. Lit. **14**, 54–62 (1976)

24. Wagner, R.E.: Viennese kaleidics: why it's liberty more than policy that calms turbulence. Rev. Austrian Econ. **25**(4), 283–287 (2012). doi:10.1007/s11138-012-0172-x
25. Barabási, A.L., Réka, A.: Emergence of scaling in random networks. Science. **286**, 509–512 (1999)
26. Granovetter, M.S.: The strength of weak ties. Am. J. Sociol. **78**(6), 1360–1380 (1973)
27. Bourdieu, P: The forms of capital. In: Richardson, J (ed) Handbook of Theory and Research for the Sociology of Education, pp. 241–258. Greenwood Press, New York (1986)
28. Coleman, J.S. Social capital in the creation of human capital. Am. J. Sociol. 94, S95–S120 (1988). Available via JSTOR. http://www.jstor.org/stable/2780243. Accessed 20 Oct 2016
29. Peteraf, M.A.: The cornerstones of competitive advantage: A resource-based view. Strateg. Manag. J. **14**(3), 179–191 (1993)
30. Barney, J.B.: Firm resources and sustained competitive advantage. J. Manag. **17**(1), 99–120 (1991)
31. Foss, K., Foss, N.J., Klein, P.G., Klein, S.K.: The entrepreneurial organization of heterogeneous capital. J. Manag. Stud. **44**(7), 1165–1186 (2007)
32. Foss, N.J.: "Austrian" determinants of economic organization in the knowledge economy. In: Birner, J., Garrouste, P. (eds) Markets, Information and Communication. Austrian Perspectives on the Internet Economy, pp. 143–168. Routledge, London (2004)
33. Chiles, T.H., Tuggle, C.S., McMullen, J.S., Bierman, L., Greening, D.W.: Dynamic creation: extending the radical Austrian approach to entrepreneurship. Organ. Stud. **31**(1), 7–46 (2010)
34. Foss, N.J., Klein, P.G., Kor, Y.Y., Mahoney, J.T.: Entrepreneurship, subjectivism, and the resource-based view: towards a new synthesis. Strateg. Entrep. J. **2**(1), 73–94 (2008)
35. Callon, M., Méadel, C., Rabeharisoa, V.: L'économie des qualities. Politix. **13**(52), 211–239 (2000)
36. Prahalad, C.K., Ramaswamy, V.: The Future of Competition: Co-creating Unique Value with Customers. Harvard Business School Press, Boston (2004)
37. Porter, M., Kramer, M.: The big idea: creating shared value; how to reinvent capitalism and unleash a wave of innovation and growth. Harv. Bus. Rev. **89**(1/2), 62–77 (2011)
38. Normann, R., Ramírez, R.: From value to value constellation: designing interactive strategy. Harv. Bus. Rev. **71**(4), 65–77 (1993)
39. Koppl, R., Mongiovi, G.: Introduction. In: Koppl, R., Mongiovi, G. (eds) Subjectivism in Economic Analysis: Essays in Memory of Ludwig M, pp. 1–11. Lachmann. Routledge, London (1998)
40. Barabasi, A.L.: Network Science. Cambridge University Press, Cambridge (2016)
41. McCloskey, D.N.: How to buy, sell, make, manage, produce, transact, consume with words. In: Clift, E.M. (ed) How Language is Used to do Business: Essays on the Rhetoric of Economics. Mellen Press, Lewiston (2008)
42. Polanyi, K.: Our obsolete market mentality: "Civilization must find a new thought pattern". Commentary. **3**(2), 109–118 (1947)
43. Winch, P.: The Idea of a Social Science and its Relation to Philosophy, 2nd edn. Routledge, London (1990)
44. Prahalad, C.K.: Ramaswamy V (2004) Co-creating unique value with customers. Strateg. Leadersh. **32**(3), 4–9 (2004)
45. Gibson, J.J.: The theory of affordances. In: Shaw, R., Bransford, J. (eds) Perceiving, Acting, and Knowing, pp. 67–82. Lawrence Erlbaum, Hillsdale (1977)
46. Gaver, W.W.: Technology affordances. In: Proceedings of the SIGCHI Conference on Human Factors in Computing Systems, New Orleans, LA, USA, April 27–May 2, 1991, pp. 79–84 (1991)
47. Barabasi, A.L.: Linked: How Everything is Connected to Everything Else and What it Means for Business, Science, and Everyday Life. Basic Book, New York (2004)

Open Questions in Multidimensional Multilevel Network Science

Jeffrey H. Johnson

Abstract Network science has made great progress in the study of binary relationships between pairs of elements. Although it has been known for decades that n-ary are ubiquitous in complex systems, progress in this area has been much slower. A condensed account is given of the family of network structures which includes graphs, networks, multilevel networks and multiplex networks for binary relations, and hypergraphs, simplices complexes and hypernetworks for n-ary relations. These structures are naturally integrated in a generalising framework. This family of network structures supports a new theory of multilevel systems where structures at one level become vertices at higher levels through part-whole aggregation interleaved with taxonomic aggregation. Although the structures presented are necessary to understand the dynamics of complex multilevel systems, there are many open questions. These are presented for consideration by the network community.

Keywords n-ary relation • Graph • Hypergraph • Network • Simplicial complex • Q-analysis • q-percolation • Multiplex network • Hypernetwork • Multilevel systems

1 Introduction

Network science has made great progress in the study of binary relationships between pairs of elements. It is now becoming more widely accepted that there is a need to embrace n-ary relations in network science [17]. This paper presents a condensed account of a family of structures able to represent n-ary relations, and the algebraic theory of multilevel systems they support. Although multidimensional

J.H. Johnson (✉)
Faculty of Science, Technology, Engineering and Mathematics, The Open University,
Milton Keynes MK7 6AA, UK
e-mail: jeff.johnson@open.ac.uk

© Springer International Publishing AG 2017
E. Shmueli et al. (eds.), *3rd International Winter School and Conference on Network Science*, Springer Proceedings in Complexity, DOI 10.1007/978-3-319-55471-6_10

network structures are necessary to understand the dynamics of complex multilevel systems, there are many open questions. These are set out in the concluding section for consideration by the network science community.

It will be assumed that the reader is familiar with graphs and networks. Multilayer and multiplex networks provide a formalism for the analysis of networks defined by many different relations [11]. A comprehensive account of multilayer and multiplex networks is given by Boccaletti et al. in [10].

A weakness of conventional network theory is that the notation $\langle v, v' \rangle$ does not discriminate between the defining relations. To do this an extra symbol is required. For example, let V represents a set of people, R_1 the relation 'is line managed by' and R_2 the relation 'plays golf with'. Then v and v' may satisfy both relations. Let the notation $\langle v, v'; R_1 \rangle$ means v' is the boss of v. This is different to $\langle v, v'; R_2 \rangle$ meaning that v plays golf with v'. This notation has the desirable feature that it naturally allows the definition of algebraic operations on the relations such as $\langle v, v'; R_1 \wedge R_2 \rangle$ which combines the relations R_1 and R_2 to form composite relations such as $(R_1 \wedge R_2)$ meaning 'plays golf with the boss'.

2 Hypergraphs and the Galois Lattice

There are many of instances n-ary relations between more than two vertices. For example, consider four people playing bridge. This is a 4-ary relation. n things are related by an n-ary relation if it ceases to hold on removing any them. For example, if one person leaves the bridge game, the game no longer continues normally. The structures at the top of Fig. 1 generalise the structures at the bottom by allowing relations between any number of vertices.

The French mathematician Claude Berge made an early attempt to generalise relational structure to many vertices through his definition of hypergraphs developed in the 1960s.

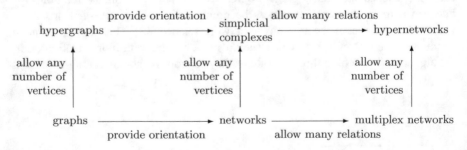

Fig. 1 The natural family of network structures embraces n-ary relations

Let $X = \{x_1, x_2, \ldots, x_n\}$ be a finite set. A *hypergraph* on X is a family $H = (E_1, E_2, \ldots, E_m)$ of subsets of X such that

(1) $E_i \neq \emptyset \qquad (i = 1, 2, \ldots, m)$
(2) $\bigcup_{i=1}^{m} = X$.

The elements $x_1, x_2, \ldots, x_m$ are called *vertices* and the sets $E_1, E_2, \ldots, E_m$ are the *edges* of the hypergraph [8, 9].

Let R be a relation between sets A and B. Let $a R b$ mean that a is R-related to b where $a \in A$ and $b \in B$. Let $R(a)$ be the set of all $b \in B$ that are R-related to a, $R(a) = \{b \mid a R b\}$. Then $H_A(B, R) = \{R(a) \mid \text{for all } a \in A\}$ is a hypergraph. Similarly, let $R(b) = \{a \mid a R b\}$. Then $H_B(A, R) = \{R(b) \mid \text{for all } b \in B\}$ is a also hypergraph.

Given the hypergraph $H_A(B, R)$, let $\mathcal{H}_A(B, R)$ be all the sets in $H_A(B, R)$ together with all their intersections. Let $\mathcal{H}_A(B, R)$ be called a *Galois hypergraph*. Similarly, let $\mathcal{H}_B(A, R)$ be all the sets in $H_B(A, R)$ together with all their intersections. Then $\mathcal{H}_A(B, R)$ and $\mathcal{H}_B(A, R)$ are *dual Galois hypergraphs*.

Proposition The sets in the dual Galois hypergraphs $\mathcal{H}_A(B, R)$ and $\mathcal{H}_B(A, R)$ are in one-to-one correspondence. This is called the *Galois connection* between the dual hypergraphs.

A proof of this proposition can be found in [18]. The intuition behind the proposition is that there are paired maximal subsets called Galois pairs, $A' \leftrightarrow B'$ where every member of $A' \subseteq A$ is R-related to every member of $B' \subseteq B$.

Proposition There is an order relation on the set of Galois pairs with an associated *Galois Lattice*

Let $A' \leftrightarrow B'$ and $A'' \leftrightarrow B''$ be Galois pairs. Then $A' \subseteq A''$ if and only if $B' \supseteq B''$. Let $\lesssim$ be defined as $A' \leftrightarrow B' \lesssim A'' \leftrightarrow B''$ if $A' \subseteq A''$. Then $\lesssim$ is an order relation with an associated lattice structure. This is called the Galois Lattice for the relation R between A and B. More details of the Galois connection and Galois Lattice can be found in [13, 14, 16, 18].

3 Simplicial Complexes and Q-Analysis

Hypergraphs have the great advantage that they are simple set-theoretic structures and this makes it easy to prove the existence of the Galois connection and Galois Lattice. However set theory is too weak for most applications because the elements are not ordered. For example, $\{R, E, P, A, I, R\} = \{R, A, P, I, E, R\}$ so the words 'repair' and 'rapier' cannot be discriminated by the sets of their letters—the *order* of the letters is also required.

At the same time that Berge was developing his theory of hypergraphs, the British mathematician Ron Atkin was developing his theory of Q-analysis based on simplicial complexes and algebraic topology [1–6].

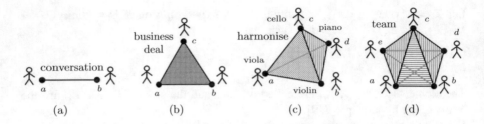

Fig. 2 Simplices can represent relations between two or more things. (**a**) Line 1-dimensional. (**b**) Triangle 2-dimensional. (**c**) Tetrahedron 3-dimensional. (**d**) 5-hedron 4-dimensional

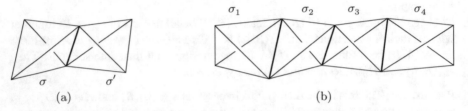

Fig. 3 q-connected polyhedra. (**a**) σ and σ' are 1-near. (**b**) σ_1 and σ_4 are 1-connected

Let V be a set of elements called *vertices*. An *abstract p-dimensional simplex* $\langle x_0, x_1, \ldots, x_p \rangle$ is an ordered set of $p+1$ vertices. A simplex $\langle x'_0, x'_1, \ldots, x'_q \rangle$ is a *q-dimensional face* of the simplex $\langle x_0, x_1, \ldots, x_p \rangle$ iff $\{x'_0, x'_1, \ldots, x'_q\} \subseteq \{x_0, x_1, \ldots, x_p\}$. A set of simplices is called a *simplicial family*. A set of simplices with all its faces is called a *simplicial complex*.

In algebraic topology it is common to use the symbol σ to represent simplices, and this convention will be used here. Simplices have a geometric realisation as p-dimensional polyhedra, as shown in Fig. 2.

Two simplices are *q-near* if they share a q-dimensional face. Two simplices are *q-connected* if there is a chain of pairwise q-near simplices between them. The tetrahedra σ and σ' are 1-near in Fig. 3a because they share a 1-dimensional face. In Fig. 3b σ_1 and σ_4 are 1-connected, since σ_1 is 1-near σ_2, σ_2 is 1-near σ_3, and σ_3 is 1-near σ_4.

A *Q-analysis determines* classes of *q-connected components*, sets of simplices that are all q-connected. An early application of Q-analysis studied land uses in Colchester [6].

4 Backcloth, Traffic and Multidimensional Percolation

The vertices and edges of networks often have numbers associated with them. For example in a social network the vertices may be associated with the amount of money a person has and the edges may be associated with how much money passes between pairs of people. In electrical networks the vertices have voltage associated with them and the edges have current. Although the network's voltages and currents may change, the network itself does not. Similarly in a road network the daily traffic flows may vary but usually the network infrastructure does not. The same holds for simplicial complexes when there are patterns of numbers across the vertices and the simplices. The numbers may change when the underlying simplicial complex does not.

Atkin suggested that the relatively unchanging network or simplicial complex structure be called a *backcloth* and that the numbers be called the *traffic* of activity on the backcloth. As an example, the airline network acts as a backcloth to the traffic of airline passengers. The term backcloth comes from the scenery painted on large canvas sheets used in theatres as a static backdrop behind the actors.

Atkin first used simplicial complexes to characterise a wide variety of phenomena in physics by his *Cocycle Law* that the space-time backcloth supporting many physical phenomena has no holes. His conceptual leap "from cohomology in physics to q-connectivity in social science" was published in 1972 [1, 12].

Networks are excellent for representing and calculating the dynamics of flows, including electricity, fluids, vehicles and sentiments. Simplicial complexes are multidimensional networks and they too can carry equally diverse traffic flows. Generally the q-connectivity of the underlying backcloth constraints the dynamics of the flows. This has been called q-*transmission* and has been described as a multidimensional analogue to percolation in networks [15, 16].

5 Hypernetworks

Although simplicial complex are a step forward in representing n-ary relations they too have their limitations, as illustrated in Fig. 4. Here the lines $\ell_1, \ldots, \ell_{16}$ are arranged in a circle by the relation R_1. The resulting structure $\langle \ell_1, \ldots, \ell_{16}; R_1 \rangle$ has the emergent property that most people see a white disk at the centre of the lines,

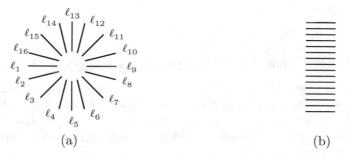

(a)   (b)

Fig. 4 The lines $\ell_1, \ldots, \ell_{16}$ organised by two different relations, R_1 and R_2. (a) The sun illusion $\sigma_1 = \langle \ell_1, \ldots, \ell_{16}; R_1 \rangle$. (b) The rectangle illusion $\sigma_2 = \langle \ell_1, \ldots, \ell_{16}; R_2 \rangle$

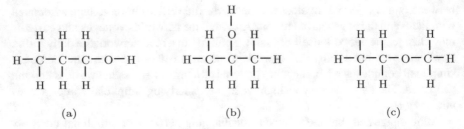

Fig. 5 Chemical isomers as relational simplices. (**a**) *n*-propyl alcohol. (**b**) Isopropyl alcohol. (**c**) Methyl-ethyl-ether

the so-called *sun illusion*. Figure 4b shows the same set of lines assembled under a different relation, R_2. Now there is no disk but a rectangle shape emerges. This example illustrates that the same ordered set of elements can be the subject of more than one relation, and that the simplex notation $\langle \ell_1, \ldots, \ell_{16} \rangle$ cannot discriminate these very different cases.

In order to do this another symbol is necessary to represent the relation. We write $R_1 : \langle \ell_1, \ldots, \ell_{16} \rangle \rightarrow \langle \ell_1, \ldots, \ell_{16}; R_1 \rangle$ and $R_2 : \langle \ell_1, \ldots, \ell_{16} \rangle \rightarrow \langle \ell_1, \ldots, \ell_{16}; R_2 \rangle$. Let σ_1 represent the sun configuration and σ_2 represent the rectangle configuration. Then σ_1 and σ_2 are examples of *relational simplices*, or *hypersimplices*. Now the notation enables σ_1 to be discriminated from σ_2, since $\sigma_1 \neq \sigma_2$.

As another example, propanol assembles three carbon atoms with eight hydrogen atoms and one oxygen atom, written as C_3H_8O or C_3H_7OH. Figure 5 shows the atoms of propanol arranged in a variety of ways. The first two show the isomers *n*-propyl alcohol and isopropyl alcohol. The oxygen atom is attached to an end carbon in the first isomer and to the centre carbon in the second, but the C-O-H hydroxyl group substructure is common to both. The rightmost isomer of C_3H_8O, methoxyethane, has the oxygen atom connected to two carbon atoms and there is no C-O-H substructure. This makes it an ether, methyl-ethyl-ether, rather than an alcohol. Thus the relational simplices of the isomers have the same vertices, but the assembly relations are different. *n*-propyl alcohol and isopropyl alcohol share the hydroxyl group substructure C-O-H and are similar, but methyl-ethyl-ether does not and has different properties. Thus

$$\langle\, C, C, C, H, H, H, H, H, H, H, H, O \,;\, R_{\,n-\text{propylalcohol}} \rangle \quad \neq$$
$$\langle\, C, C, C, H, H, H, H, H, H, H, H, O \,;\, R_{\,\text{isopropylalcohol}} \rangle \quad \neq$$
$$\langle\, C, C, C, H, H, H, H, H, H, H, H, O \,;\, R_{\,\text{methyl}-\text{ethyl}-\text{ether}} \rangle$$

In general a *hypernetwork* is defined to be any collection of hypersimplices. This definition is deliberately undemanding, so that almost anything can be a hypersimplex, and any collection of hypersimplices can be a hypernetwork. Hypersimplices can act as backcloth structure carrying a traffic of numbers on their vertices and on their faces.

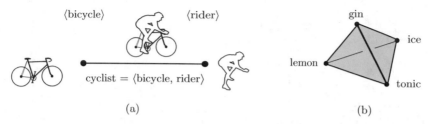

(a)   (b)

Fig. 6 Remove a vertex and the simplex ceases to exist. (**a**) Remove a vertex and the cyclist simplex ceases to exist. (**b**) Remove a vertex and the perfect gin and tonic ceases to exist

6 The Vertex Removal Test for *n*-ary Relations

The essential feature of a polyhedron is that it ceases to exist if any of the vertices are removed. For example, consider a cyclist represented as the combination ⟨rider, bicycle; R_{riding}⟩. Remove either the man or the bicycle and what is left ceases to be a cyclist. Remove any vertex from ⟨gin, tonic, ice, lemon; R_{mixed}⟩ and it ceases to be the perfect gin and tonic (Fig. 6). Generalising edges to polyhedra allows a distinction to be made between the *parts* of things represented by vertices, and *wholes* represented by hypersimplices. Using this test it is easy to find many examples of *n*-ary relations, e.g. a path with *n* edges in a network forms a hypersimplex—remove an edge and the path ceases to exist; four bridge players form a hypersimplex—remove one and the game collapses; and a car and its wheels are 5-ary related—without any of them it won't work.

7 Hypernetworks and Multilevel Structure

Hypersimplices enable the definition of multilevel part-whole structures, e.g. the four blocks assembled by the 4-ary relation *R* to form an arch in Fig. 7. Here the whole has the emergent property of a gap not possessed by any of its parts. If the parts exist in the system at an arbitrary *Level N* then the whole exists at a higher level, here shown as *Level N+1*. Thus assembly relations provide an immutable upwards arrow for the definition of multilevel structure.

Part-whole aggregations are interleaved with taxonomic aggregations, as shown in Fig. 8. The aggregation between *Level N* and *Level N+1* combines graphical parts to form faces. The aggregation between *Level N+1* and *Level N+2* establishes classes of faces in a taxonomy. Such aggregations depend on the purpose of the taxonomy. For example, there is no class of 'frowny' faces because, for the purpose here, it is not required. Note that part-whole aggregations require *all* the parts. In contrast taxonomic aggregations require just one example to aggregate. For example, the round smiley face is sufficient for there to be a smiley face, irrespective of whether or not there is a square smiley face.

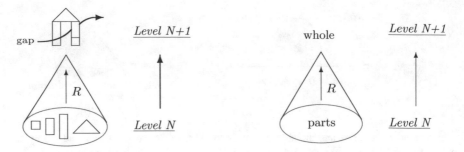

Fig. 7 The fundamental part-whole diagram of multilevel aggregation

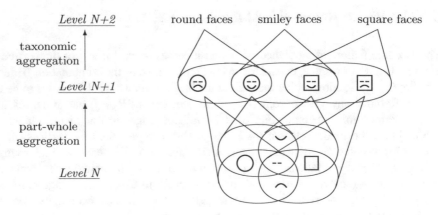

Fig. 8 Part-whole and taxonomic aggregation

8   The Multilevel Fragment-Recombine Operator

When dealing with multilevel systems it would be useful to have a single symbol to represent the very complicated multilevel cone structures illustrated in Fig. 9a. One possibility is to enclose them by triangle. This representation allows a subsystem to be represented by a triangle within a triangle as shown in Fig. 9b. Since the intersection of two triangles is also a triangle, this representation is convenient to denote the intersection of two multilevel systems, as shown in Fig. 9c.

This representation suggests an exciting new possibility for multilevel complex systems. To be more concrete consider a narrative as a multilevel structure made of words, phrases, paragraphs and complete stories. Narratives are very important in policy and very important for the development of a theory of complex social systems.

For example, Europe is grappling with many narratives associated with migrants, and these narratives work at the level of the plight of individual people, through to more aggregate structures such as people traffickers' boats to more aggregate structures such as countries and their policies. The narratives include political and

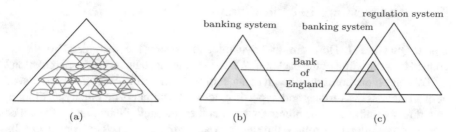

Fig. 9 Multilevel operations on multilevel triangles. (**a**) A multilevel triangle. (**b**) Subsystem.
(**c**) Intersection

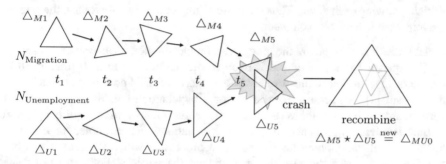

Fig. 10 Multilevel fragment-recombine operators

economic aspects at many level of aggregation. Let this multilevel narrative be
called $N_{\text{Migration}}$ as shown top-left in Fig. 10. Let $\triangle_{Mi}$ be the *state* of the narrative
at time t_i.

Alongside the strong migration narrative in the UK there are others, e.g. the
unemployment narrative, $N_{\text{Unemployment}}$, shown bottom-left in Fig. 10.

Both these narratives evolve in time, with information and invention being added
or lost as the meanings of the narrative evolve. Figure 10 shows these narratives
evolving *independently* until they *crash* into each other at time t_5. The combinatorial
dynamics of such a crash is not well understood, but it involves parts of the two
multilevel systems interacting and each of the multilevel narratives fragmenting
before they *recombine* to form new *composite* narratives, e.g. $N_{\text{migrants are taking our jobs}}$.
Let the *fragment-recombine operator* of multilevel systems, $\star$, be defined as

$$\star : (\triangle_1, \triangle_2) \rightarrow \triangle_1 \star \triangle_2$$

where $\triangle_1$ and $\triangle_2$ are multilevel systems before they crash and $\triangle_1 \star \triangle_2$ is the
multilevel system after. There are many open questions associated with this.

9   Open Questions in Multidimensional Network Science

Open Question 1 How are the dynamics of systems constrained by the q-connectivity structure of the backcloth? What are the mechanisms for q-percolation?

This question concerns the dynamics of system traffic, *i.e.* the patterns of numbers across the vertices and hypersimplices. The numbers on one hypersimplex can directly influence the numbers on another through their shared face. For example, consider two routes through a road traffic system. The routes can be considered to be structured sets of road segments, $R_i = \langle s_0, s_1, \ldots, s_n; R_{\text{route}} \rangle$. The more segments that two routes share, the more their traffic will interact, with the vehicles slowing each down. Thus the more highly connected the routes, the greater the impact they have on each other.

Open Question 2 What are the processes of hypersimplex formation and loss?

This question is a generalised version of the question as to how links form in networks. One answer to this for networks is the Barabási's preferential attachment mechanism [7]. A necessary condition for a hypersimplex to form is that all its vertices are present. For the vertices to become n-ary related may require a process in time involving a sequence of other p-ary relations. For example, for n people to form a well-functioning team involves a process in which they learn to work together. In the case the process may change the vertices. For example, some members of the team may be trained in order to acquire necessary skills.

Open Question 3 What is the nature of multilevel backcloth-traffic dynamics?

This question combines the first and second questions in the context of multilevel interactions. Bottom-up the traffic or patterns of numbers aggregate over the multilevel backcloth. In general more aggregate numbers get larger and more predictable in time. Furthermore, aggregation bottom-up tends to convert lower level structures into numbers. For example, company taxation results in a time series of numbers at the national level, where the details of companies and their activities are not explicit. Similarly there are issues of how numbers are distributed top-down over multilevel systems. The challenge is to understand how bottom-up and top-down dynamics interact across multilevel systems.

Another aspect of this question concerns the formation of multilevel structure, and the processes by which top-down decisions enable or require the creation of lower level structures. For example a company may decide to invest in a new factory. This requires information traffic to flow across higher management level resulting in top-down flows of resources to create the lower level structure. Generally the rationale for this is that the new lower structure will create new resources that will flow up the structure to enhance the company's profits.

When modelling systems it is always the case that some things are included and some are left out. This includes deciding that some level is the lowest necessary to model a multilevel system. The dynamics of a multilevel system are said to be *information complete* at *Level N* if modelling their behaviour requires no information from *Level N-k* for $k \geq 1$. Thus Open Question 3 includes how to decide the level at which a system is information complete. For example, until recently, economic systems were modelled at the meso level of the 'representative agent'. Today it is increasing realised that many social system are information-complete only at the level of the individual person. For example, the behaviour of road traffic system emerges from the decisions of individual driver agents and increasingly they are modelled at this level by agent-based simulations using the disaggregate data of synthetic micro populations.

Open Question 4 What new algebraic operations can be defined between hyper-simplices?

The generality of this question is given by the expression

$$\langle x_0, \ldots, x_{p_1}; R_1 \rangle \diamond \langle y_0, \ldots, y_{p_2}; R_2 \rangle = \langle z_0, \ldots, z_{p_{1,2}}; R_1 \oslash R_2 \rangle$$

where $\{z_0, \ldots, z_{p_{1,2}}\} = \{x_0, \ldots, x_{p_1}\}\boxed{?}\{y_0, \ldots, y_{p_2}\}$. The challenge is to determine the nature of the operators $\boxed{?}$ and $\oslash$.

This question has its origins in the simple question "what is the intersection of two hypersimplices?" An obvious but unsatisfactory answer is given by

$$\langle x_1, x_2, \ldots, x_n; R_1 \rangle \cap \langle y_1, y_2, \ldots, y_{p'}; R_2 \rangle \stackrel{?}{=} \langle z_1, z_2, \ldots, z_q; R_1 \wedge R_2 \rangle,$$

where $\{z_1, z_2, \ldots, z_q\} = \{x_1, x_2, \ldots, x_p\} \cap \{y_1, y_2, \ldots, y_{p'}\}$.

The problem here is that R_1 is defined on *all* the vertices $x_1, x_2, \ldots, x_p$ and R_2 is defined on *all* the vertices $y_1, y_2, \ldots, y_{p'}$ but, as written, their conjunction is defined on $\{x_1, x_2, \ldots, x_p\} \cap \{y_1, y_2, \ldots, y_{p'}\}$. Another possibility is to write

$$\langle x_1, x_2, \ldots, x_n; R_1 \rangle \cap \langle y_1, y_2, \ldots, y_{p'}; R_2 \rangle \stackrel{?}{=} \langle z_1, z_2, \ldots, z_q; R_1 \wedge R_2 \rangle,$$

where $\{z_1, z_2, \ldots, z_q\} = \{x_1, x_2, \ldots, x_p\} \cup \{y_1, y_2, \ldots, y_{p'}\}$.

To investigate this question further consider two multiplex network edges, $\langle x_1, x_2; R \rangle$ and $\langle y_1, y_2; R' \rangle$

$$\langle x_1, x_2; R \rangle \cap \langle y_1, y_2; R' \rangle \stackrel{\text{def}}{=} \emptyset \text{ for } \{x_1, x_2\} \cap \{y_1, y_2\} = \emptyset$$

$$\stackrel{\text{def}}{=} \langle x_1, x_2; R \wedge R' \rangle \text{ for } \{x_1, x_2\} \cap \{y_1, y_2\} = \{x_1, x_2\}$$

$$\stackrel{\text{def}}{=} \langle x_1, x_2 \rangle \boxed{?} \langle y_1, y_2 \rangle; R \oslash R' \rangle \text{ otherwise}$$

Of these, the first and second are not problematic, the former being an empty intersection and the latter being the conjunction of the relations. But how should the $\boxed{?}$ and $\oslash$ operations be defined?

Suppose $\langle x_1, x_2 \rangle \boxed{?} \langle y_1, y_2 \rangle = \langle x_1, x_2 \rangle \cap \langle y_1, y_2 \rangle = \langle x_1 \rangle$. This means that $R \oslash R'$ is defined on a single vertex. It is perhaps more promising to suppose that $\langle x_1, x_2 \rangle \boxed{?} \langle y_1, y_2 \rangle = \langle x_1, x_2 \rangle \cup \langle y_1, y_2 \rangle$?

Open Question 5 What is the nature of the multilevel fragment-recombine operator $\star : (\triangle_1, \triangle_2, \ldots) \rightarrow \triangle_1 \star \triangle_2 \star \ldots$ for multilevel systems.

It may be easiest to answer by thinking ahead to how the $\star$ operation could be implemented. In practice it is assumed that the multilevel systems are explicitly represented in multilevel data structures based on the algebra sketched in this paper. Then the question becomes how hypersimplices at compatible levels behave when they crash into each other. Presumably there are various $\boxed{?}$ and $\textcircled{?}$ operations to deconstruct and recombine the hypersimplices. The nature of these is a major challenge for multilevel multidimensional network science.

Acknowledgement Supported by the UK Home Office and HEFCE through a Police Knowledge Fund grant to the Open University National Centre for Policing Research and Professional Development.

References

1. Atkin, R.H.: From cohomology in physics to Q-connectivity in social science. Int. J. Man Mach. Stud. **4**(2), 139–167 (1972)
2. Atkin, R.H.: Mathematical Structure in Human Affairs. Heinemann Educational Books, London (1974)
3. Atkin, R.H.: Combinatorial Connectivities in Social Systems. Birkhäuser, Basel (1977)
4. Atkin, R.H.: Multidimensional Man. Penguin Books, Harmondsworth (1981)
5. Atkin, R.H., Bray, R., Cook, I.: A mathematical approach towards a social science. The Essex Review, University of Essex, Autumn 1968, No. 2, 3–5 (1968)
6. Atkin, R.H., Johnson, J.H., Mancini, V.: An analysis of urban structure using concepts of algebraic topology. Urban Stud. **8**, 221–242 (1971)
7. Barabási, A.-L.: Linked. Perseus Books Group, Cambridge (2002)
8. Berge, C.: Hypergraphs: Combinatorics of Finite Sets. North Holland, Amsterdam (1989)
9. Berge, C.: Sur certains hypergraphes généralisant les graphes bipartites. In: Erdös, P., Rhényi, A., Sós, V.T. (eds.) Combinatorial Theory and Its Applications I. Proceedings of the Colloquium on Combinatorial Theory and Its Applications, 1969, pp. 119–133. North-Holland, Amsterdam (1970)
10. Boccaletti, S., Bianconi, G., Criado, R., del Genio, C.I., Gómez-Gardeñes, J., Romance, M., Sendiña-Nadal, I., Wang, Z., Zanin, M.: The structure and dynamics of multilayer networks. Phys. Rep. **544**, 1–122 (2014)
11. De Domenico, M., Solé-Ribalta, A., Cozzo, E., Kivela, M., Moreno, Y., Porter, M.A., Gómez, S., Arenas, A.: Mathematical formulation of multilayer networks. Phys. Rev. X **3**, 041022 (2013). http://journals.aps.org/prx/pdf/10.1103/PhysRevX.3.041022
12. Dowker, C.H.: The homology groups of relations. Ann. Math. **56**(1), 84–95 (1952)
13. Freeman, L.C., White, D.R.: Using Galois lattices to represent network data. In: Sociological Methodology, vol. 23. American Sociological Association, Washington (1993). ISBN 1-55786-464-0, ISSN 0081–1750 http://eclectic.ss.uci.edu/$\sim$drwhite/pw/Galois.pdf
14. Freeman, L.C., White, D.R., Romney, A.K.: Research Methods in Social Network Analysis. Transaction, New Brunswick (1991)
15. Johnson, J.H.: Hypernetworks for reconstructing the dynamics of multilevel systems. In: European Conference on Complex Systems 2006, Oxford, 25–29 September 2006. http://oro.open.ac.uk/4628/1/ECCS06-Johnson-R.pdf

16. Johnson, J.H.: Hypernetworks in the Science of Complex Systems. Imperial College Press, London (2014)
17. Johnson, J.H.: Embracing n-ary relations in network science. In: Wierzbicki, A., Brandes, U., Schweitzer, F., Pedreschi, D. (eds.) Proceedings of 12th International Conference and School on Advances in Network Science, NetSci-X 2016, Wroclaw, 11–13 January 2016
18. Johnson, J.H.: Hypernetworks: multidimensional relationships in multilevel systems. Eur. Phys. J. Spec. Top. **225**(6–7), 1037–1052 (2016). https://www.researchgate.net/publication/308956954_Hypernetworks_Multidimensional_relationships_in_multilevel_systems

Index

© Springer International Publishing AG 2017
E. Shmueli et al. (eds.), *3rd International Winter School and Conference on Network Science*, Springer Proceedings in Complexity, DOI 10.1007/978-3-319-55471-6

Printed in the United States
By Bookmasters